IMAGES
of America

WASHINGTON'S CRANBERRY COAST

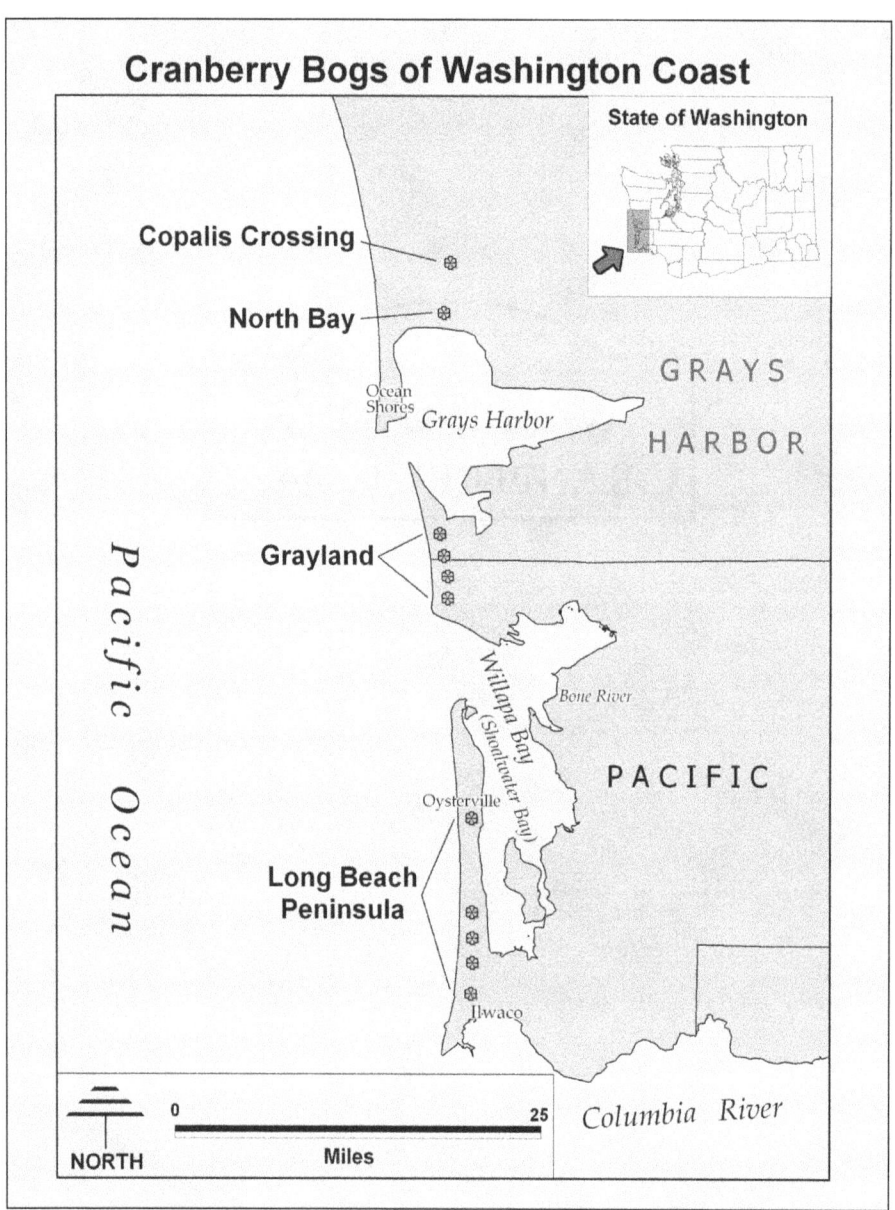

MAP OF WASHINGTON'S CRANBERRY BOGS. For untold centuries, indigenous peoples prized the bright red *pil ollalies* (Chinook Jargon for "cranberries") that grew along the seacoast north of the Columbia River. Today, domesticated cranberries are grown in the same boggy areas, stretching northward from the Long Beach Peninsula to Copalis Crossing. This is one of the few areas on earth where the colorful fruit thrives. (Paul E. Staub.)

ON THE COVER: In early years, all cranberries were harvested by hand. Pickers filled 12½ pound containers called "hampers" and were paid with round metal coins, which could be redeemed at local stores for an equivalent's worth of food or other goods. The average picker could fill eight hampers a day. In the foreground, cranberries are transferred onto drying trays, which will be warehoused to await marketing. (Christine Cottrell.)

IMAGES
of America

WASHINGTON'S CRANBERRY COAST

Sydney Stevens for the Pacific Coast
Cranberry Research Foundation
Foreword by Kim Patten, PhD

Copyright © 2018 by Sydney Stevens for the Pacific Coast Cranberry Research Foundation
ISBN 978-1-4671-2989-3

Published by Arcadia Publishing
Charleston, South Carolina

Library of Congress Control Number: 2018942547

For all general information, please contact Arcadia Publishing:
Telephone 843-853-2070
Fax 843-853-0044
E-mail sales@arcadiapublishing.com
For customer service and orders:
Toll-Free 1-888-313-2665

Visit us on the Internet at www.arcadiapublishing.com

To the many cranberry farmers, past and present, who have brightened Washington's coast each harvest and who have blessed us all with a downright saucy place in which to work and to live!

Contents

Foreword		6
Acknowledgments		7
Introduction		8
1.	A Tenuous Beginning	11
2.	Planting and Growing	29
3.	Tending and Harvesting	51
4.	Processing and Marketing	91
5.	A Thriving Community	109
Bibliography		126
Index		127

Foreword

For the past 28 years, I have had the opportunity to work with the finest group of individuals I have ever come to know: Washington's cranberry growers. Their love and passion for their land and community, and their grit and entrepreneurism, have been inspiring. These farmers have struggled against unending challenges: pests, low-producing cranberry vines, unforgiving weather, unending new regulatory permits, market oversupply, and subsequent low crop prices.

As an extension professor for Washington State University (WSU), my mission has been to help develop and implement practical solutions to many of these problems. Achieving success has been an interesting challenge. Because of budget cuts at WSU, the Coastal Washington Research and Extension Center (CWREC) was proposed for closure soon after I arrived. Thankfully, the farmers were able to prevail against the closure by forming the Pacific Coast Cranberry Research Foundation, purchasing the property and leasing it back to the university for free.

With the Research Station (CWREC) no longer in jeopardy, we were able to continue to fulfill the land-grant university's mission, which is to provide agricultural and technical education that addresses grower needs. For example, we developed new methods to control recalcitrant weeds and troublesome insects like black-vine weevil and black-headed fireworm. We also tested and released several hardier, higher-yielding cranberry varieties that would grow well in this area. These new methods to control pests and develop new varieties can mean the difference between farming at a profit or, literally, losing the farm.

It has been a pleasure to be in a long-term collaborative partnership with these coastal cranberry growers, and I hope that our efforts will help assure that many more generations of family farmers will thrive. *Washington's Cranberry Coast* clearly portrays the fascinating story of the past 130 years of these remarkable farmers and their undaunted spirit. They truly are "the salt of the earth."

—Kim Patten, PhD, WSU Extension Agent

ACKNOWLEDGMENTS

Washington's Cranberry Coast is a collaborative effort. It could not have happened without the generosity of the Pacific Coast Cranberry Research Foundation and to all those listed here.

Very special thanks go to Melinda Crowley, Cranberry Museum curator, for her careful instruction and infinite patience in guiding me through the Cranberry Museum archives. Thanks, too, go to Ardell McPhail, who allowed me to piggy-back on her years of friendship and rapport with growers up and down the coast. Also, my thanks are extended to Lee Crowley and to Malcolm McPhail, cranberry growers extraordinaire. They saturated me with information, quirky facts, and serious concerns about fragile farming operations that the public seldom sees. And also, thanks go to the go-to guy for us all: Kim Patten, Washington State University Extension Agent.

The following people contributed special assistance and information: Erin Glenn, Ruth Harpe, Shanea Harpe, Lonny Hatton, Linda Karjala, Martin Paulson, and Florence Shawa.

Those who contributed images are noted in the photographic credits as follows: Sam Beebe, Ecotrust (SBeebe); David Bellamy (DBellamy); Jean Bogh (JBogh); Bob Coppin (BCoppin); Christine Cottrell (CCottrell); Lee Crowley (LCrowley); Melinda Crowley (MCrowley); Allen Devlin (ADevlin); Rick Horn (RHorn); Charlotte Jacobs (CJacobs); Peter Krebs (PKrebs); Ardell McPhail (AMcPhail); Lane Sacks (LSacks); Kathleen Sayce (KSayce); Gordon Schoewe (GSchoewe); Karen Snyder (KSnyder); Nyel Stevens (NStevens); Sharon and Glen Thompson (S/GThompson); Jean Tweedy (JTweedy); Tucker Wachsmuth (TWachsmuth); and Clarence and Joann Warness (C/JWarness).

These institutions and groups also contributed images: Coastal Washington Research & Extension Center (CWREC); Columbia Pacific Heritage Museum (CPHM); Cranguyma Farms (CGM); Furford Museum (FM); "Graves Studio, 1921, from the Ivan Shirrod Collection, WSU Libraries' MASC" (WSULibraries); Oakland Public Library, Oakland History Room (OPL); Pacific Coast Cranberry Research Foundation (PCCRF); Pacific County Historical Society (PCHS); the *Tribune*/Murfin Family Collection (T/MFC); and University of Washington Libraries, Special Collections (UW/NA1412).

Special thanks go to growers in North Bay and Grayland who enthusiastically dropped everything (twice) to meet with us and to Connie Allen, who volunteered to act as technical assistant in Grayland and helped me understand the considerable differences between Grayland's cranberry practices and those of other localities along the coast.

And for keeping us oriented throughout, a big thanks goes to cartographer Paul Staub.

INTRODUCTION

"Persistence and resilience," say the cranberry growers of Washington coast. "Those are the most important survival qualities for farmers in this industry." Certainly, they are the enduring characteristics that have been exhibited by determined cranberry farmers for almost 150 years.

As early as 1852, first settlers along the remote western edge of the continent saw the possibilities for developing a commercial cranberry industry. They watched (and sometimes participated in) the wild cranberry (*Vaccinium oxycoccos*) harvest by the Native Americans who lived in the area. For countless generations, these indigenous peoples had harvested the bright red berries that grew in the marshy swales and peaty wetlands near the coast. They called them *pil ollalies* and used them as a winter staple and as an important trade item with inland tribes and with the sailing ships that sometimes made landfall nearby. A few early settlers even tried to transplant and domesticate the wild cranberry.

Although those first attempts at farming the wild cranberry proved unsuccessful, the idea persisted, especially among new arrivals from Massachusetts, where a thriving cranberry cultivation had been in progress since the early 1800s. It seemed logical that a similar farming operation could be developed on the Pacific coast. However, it was not until 1882 that the first successful bog was developed in Washington Territory, and even then, the difficulties encountered often limited forward progress.

Of greatest concern was the weather, so different from that of other areas of the country with long-established bogs. The unpredictable killing frosts of spring, right at the time of fruit-set, was a constant challenge. So, too, were the various pests and mildews that attacked vines and berries, to say nothing of the difficulties in hiring reliable workers to do the intensive year-round agricultural chores required. Marketing was also difficult from the remote coast. Transportation was largely by sea and, because the cranberry harvest coincided with the onset of October storms, shipping the fragile fruit to market was always a dubious proposition.

Nevertheless, as the coast became more populated, the acres of marshland with their potential caught the attention of real estate developers and syndicates. Extravagant promises were made to potential buyers: the land would be developed and planted, the bogs would be taken care of, and crops gathered and marketed, if required. Such assurances, however, did not magically solve the ongoing problems, and the fledgling industry continued to struggle. It was the persistence of a few growers determined to find help that resulted in the first glimmer of hope for the industry.

In 1922, D.J. Crowley, a candidate for a bachelor of science degree in plant pathology, was assigned by the State College of Washington to investigate the problems of the cranberry industry on the Washington coast. Almost immediately, things began to improve and, for the next 30 years, Crowley worked closely with farmers to find solutions to their problems. It was under Crowley's direction that the Cranberry Research Station (now called the Coastal Washington Research and Extension Center) was built. With its adjacent experimental bogs, the "Research Station," as it is referred to locally, continues to be the epicenter for cutting-edge solutions to West Coast cranberry problems.

The first big challenge was tackled immediately by Crowley. He arrived in Long Beach realizing that the coastal weather was far different from that in other cranberry-growing regions, and he determined to find a workable solution to the killer frost that crippled so many bogs each spring. He experimented with a number of minimally successful remedies, but it was not until he thought back to an early physics class that he found an answer. His workable solution would benefit not only Washington's cranberry farmers but also fruit growers throughout the world.

Each of Crowley's successors, in his or her turn, pursued various issues that plagued cranberry growers. Dr. Charles Doughty worked on twig blight fungus, which nearly wiped out many coastal growers. It was during this period of time that public sentiment towards pesticides made it necessary for work previously accomplished on herbicides to be redone and Doughty's successor, Dr. Azmi Shawa, in addition to working on cranberry fertility and color enhancement, devoted much of his attention to insecticide and herbicide research.

Gradually, too, other pervading problems have resolved themselves. On the Long Beach Peninsula, settlement of the area brought more full-time residents. This, in turn, resulted in fewer absentee or part-time bog owners, and the cranberry farming community began to stabilize. In Grayland, when increased property sales threatened to cut bog owners off from their bogs, state, county, and city cooperation with the farmers resulted in a workable solution. A cooperative agreement with Ocean Spray helped with marketing issues, and the cranberry industry of Washington took on an air of permanence.

By the time Dr. Kim Patten became the extension agent in 1990, WSU Research Station at Long Beach was well-established. Its contributions to the industry extended well beyond coastal Washington, with growers in Oregon, New Jersey, Wisconsin, and Massachusetts, as well as in the Canadian provinces of British Columbia and Nova Scotia benefitting from the work done there.

Soon after Patten arrived, however, Washington State University announced its decision to close CWREC and discontinue the local extension agent's position. Farmers were outraged. Fully aware of the historical struggles of the industry and with a deep first-hand knowledge of the benefits provided by a working research partner, they developed a plan to save the station and the research position.

Years of close cooperation on farming matters stood growers in good stead as they worked toward a solution to this serious situation. Within a few months, they applied for nonprofit status and research grants, invested in income-producing property, and worked up a memorandum of understanding to present to the decision-makers of Washington State University. As a result of their hard work and careful preparation, the outcome of that fateful meeting was a win-win situation for all parties involved. The establishment of the Pacific Coast Cranberry Research Foundation, unique in the annals of the cranberry industry, is a continuing success story for the growers of the Washington coast.

THE CRANE-BERRY. Because the pink cranberry blossom resembles a crane's head, early settlers named the plant "crane-berry." Over time, the name has morphed into the now-familiar cranberry, and its bright red berries have become America's harbinger of fall. Egrets and great blue herons, similar in appearance to cranes, can often be seen near the bogs on Washington's coast—a reminder of the cranberry's original name. (GSchoewe.)

One

A Tenuous Beginning

CRANBERRIES! The cranberry plant is a low-growing, evergreen vine whose trailing stems weave themselves into a thick mat over the peaty earth in which it thrives. The fruit, a berry larger than the leaves, is nutritious but acidic in taste. One of only three fruits native to North America, cranberries are grown domestically in eastern Canada, New England, Wisconsin, and the Pacific Northwest. (AMcPhail.)

SOUGHT-AFTER STAPLE. Indigenous peoples of the western coast gathered wild cranberries each fall. The bright red berries were dried, crushed, and mixed with fat and deer meat or salmon to make a rich paste, which was then formed into cakes. These cranberry patties were used as a portable staple on journeys and considered important trade items with tribes upriver as well as with the early explorers and settlers. The berries were also prized for their medicinal purposes, and their red juice was used by Chinookan women for face paint and to dye basketmaking materials. As early as the 16th century, trading vessels such as the *Lady Washington* (1989 replica seen here) on the long return voyages from the Far East put in along the Washington coast for cranberries, as their high vitamin C content was a much-needed protection against scurvy. (Above, RHorn; left, PCHS.)

EARLY MARKETING EXPERIMENT. James G. Swan (shown here with his Haida friend Jonny Kit Elswa in 1883) arrived in Shoalwater Bay country in 1852. He was variously an oysterman, customs inspector, secretary to congressional delegate Isaac Stevens, journalist, reservation schoolteacher, lawyer, judge, school superintendent, railroad promoter, natural historian, and ethnographer. Above all, Swan was a chronicler. He wrote *The Northwest Coast or Three Years Residence in Washington Territory*, one of the earliest books describing life among the shallow tide pools and quiet forests of the Pacific coast. "The cranberry which is very plentiful," he wrote, "forms quite an article of traffic between whites and Indians." The berries reminded him of the cranberries in his native Massachusetts, and he experimented in transplanting them to his acreage on the Bone River on the east side of Shoalwater Bay. After harvesting his first crop, Swan tried sending them to market in San Francisco, but his experiment did not work out. (UW/NA1412.)

Built at Coos Bay, Oregon, 1868 — LOUISA MO[RGAN]
For Morgan & Co., Oyster Dealers.
San Francisco.

WEST COAST ISOLATION. Though Washington Territory was created in 1853 and donation land claims were offered by the federal government as an enticement to settlement, the coastal regions were difficult to reach and remained isolated for many years. There was some overland traffic due to "spillover" from the Oregon Trail, but the main transportation routes were by water. Intrepid pioneers in Washington's oyster industry established the first commercial trade routes between Shoalwater

Bay and San Francisco, and it was on oyster schooners such as the *Louisa Morrison* that marketable goods (including cranberries) were first transported between the remote north coast and the bustling markets of California. It would not be until the transcontinental railroad was completed in 1869 that the coastal regions began to seem more accessible. (TWachsmuth.)

CAPE COD WEST. James Swan was not the only early settler who found the Long Beach Peninsula reminiscent of New England. Not only the wild cranberries growing in the peaty soil of the marshes but the "whole feel of the place" was often likened to Cape Cod, Massachusetts. The Peninsula, like Cape Cod country, comprises an area of sandy elevations and swamps interspersed with small lakes and it is in these marshes in the sandy regions that the first trials at domesticating the wild cranberries of Washington were attempted. Today, bogs of domestic berries form a colorful patchwork throughout much of the south end of the Peninsula. Some plantings are more than 80 years old and are still producing well. (Both, AMcPhail.)

CLEARING THE LAND. John Marshall, a carpenter by trade, arrived in Oysterville in 1863. Marshall was from New York, where "times were dull," and had come to the West Coast hoping for work. Clearly, he was not prepared for the difficulties he met. In a letter to his wife, Marshall writes, "I am getting out timber to build a house for John Morgan. I go in the woods and look for trees suitable for house but can't find them they are so bige [sic]. They are from 150 feet to 200 feet high and from 3 feet through to 6 feet and it is so much labor to get them they stand so close together that we can't hardly get through them. I never saw such woods until I came to this country. Here we can't see anything else. People here own claims of 100 acres and 300 and can't get land enough cleared to raise a few potatoes to eat. They have to ship them it is so hard to clear off the land." (KSayce.)

GRADUAL SETTLEMENT. By 1870, according to the federal census, there were 738 people living in Pacific County. More than half were men. Some worked in the timber industry or in mills, some were farmers, and some were fishermen. But most were oystermen and, perhaps, it was for that reason that Oysterville on Shoalwater Bay had been chosen as the county seat in 1855. Because the major occupations involved taking salmon from the Columbia, oysters from Shoalwater Bay, and the timber from the Willapa Hills to supply the burgeoning markets of California, the next two decades are sometimes called "the Age of Extraction." Farming was mostly a subsistence occupation, but it would not be long before some visionary pioneers saw the possibilities for cranberries. (Above, PCHS; below, NStevens.)

AN UNFORTUNATE EXPERIENCE. John Peter Paul was born in Ohio in 1828. Though trained as a carpenter, he came to the Long Beach Peninsula in 1869 specifically to cultivate wild cranberries. He kept meticulous records on handmade shakes. He recorded the weather, measured rainfall and growth rates, and noted yearly yields. Several stories are told of the fate of those records. According to one version, Paul entrusted them to his neighbor for safekeeping, but they were lost in a house fire. Another story is that they were stored in an abandoned cabin that eventually fell in, and the records were "lost to history." In 1875, Paul moved to Oysterville, built the first Pacific County Courthouse and a new two-story school for Oysterville District No. 1. He is remembered as a master carpenter—not as an experimental cranberry farmer. (Both, TWachsmuth.)

THE FIRST BOG. Anthony Chabot holds the distinction of making the first successful planting of domestic cranberries on the West Coast. In early 1880, his East Coast brother-in-law convinced Chabot that the Long Beach Peninsula's peaty soil could be successfully adapted to the cultivation of commercial cranberries. Chabot, a water engineer in California, purchased 1,200 acres of government land on the Peninsula and ordered vines of the McFarlin variety from Massachusetts. The vines were baled and shipped around the Horn to the Peninsula but, unbeknownst to Chabot or to the workers he hired, the vines were full of pests and mildew and were unable to withstand the vagaries of northwest frost. Chabot put his nephew Robert in charge as overseer of his company, and 35 acres were put into cultivation. Despite the many challenges, including finding adequate and reliable workers, the first harvest was made in the fall of 1883, and the Chabot bog struggled forward. (Above, PCCRF; below, T/MFC.)

ANTHONY CHABOT. By the late 1880s, Chabot's production had reached an acceptable barrels-per-acre amount and his labor difficulties had apparently been solved. At harvest, local Indians, Chinese under contract, women, and children on work release from school did the picking. Many old-timers in the late 20th century remembered that their parents and grandparents had worked at the Chabot bog. But the good fortune did not last. In 1888, Anthony Chabot died, and four years later, Robert Chabot left his uncle's farm to begin bogs of his own in Grays Harbor County. Without an experienced grower in charge, Anthony Chabot's bog suffered, and by the early 1900s, it had succumbed to weeds and was essentially abandoned. History remains silent concerning Robert Chabot's reasons for leaving his uncle's cranberry enterprise. The only clue regarding his Peninsula departure is a remembered story that he had found suitable peat soil in Grays Harbor County and decided to begin his own bog. His farm at Copalis Crossing was the first of the cranberry bogs in that area. (OPL.)

EARLY CRANBERRY FARMERS. From 1883 to 1910, there were very few cranberry growers living on the Peninsula. After the Chabot company went out of business in 1904, there were only four producers still in business. Typically, farms were small and cranberries were not the farmers' main source of income. Many problems needed to be overcome, including the limited seasonal demand for cranberries; the cost of marketing and distribution because of the Peninsula's isolation; the large initial investment required, which could not be recouped for four or five years; the difficulty in importing healthy vines from the East Coast; the profits that were often unequal to the taxes they were paying for farmland; and the complications inherent in absentee ownership. Perhaps this last reason was the most troubling of all. Unable to tend their own bogs or to properly oversee hired workers, owners abandoned their farms to weeds. It was easier on their pocketbooks to let the bogs revert to swamp and marsh. (Above, CPHM; below, PCCRF.)

STAGNANT DEVELOPMENT BY 1900. Although a few farmers persevered, grubbing out a smattering of acres at a time, most development on the Long Beach Peninsula stagnated in the first decades of the 20th century. Bog owners were hard pressed to sell their property no matter how many improvements they had managed to make, and hundreds of acres up the center of the Peninsula became a financial liability. In the early 1900s, several marsh owners were momentarily encouraged when speculators proposed that the peat could be mined and dried for fuel. This seemed, for a brief time, to be the perfect sales pitch to potential buyers. However, with the abundance of cheap wood available in the Northwest, it soon became clear that the peat briquette manufacturing business would not work, and the idea had faded from the scene by 1907. (Above, KSayce; below, JBogh.)

NORTH IN GRAYLAND. The cranberry development begun in 1892 in Copalis by Robert Chabot captured the interest of other Grays Harbor County residents, including a group of Finnish immigrants. In 1912, attracted to the berry so similar to their native lingonberry, five Finnish farmers established bogs in an area soon known as "Little Finland." Their original bogs were five-acre tracts, generally 150 feet wide by 1,280 feet long and covered an area about three and one-half miles long by a half-mile wide. By 1938, they had approximately 250 acres in production. Because of the configuration of the bogs—long, narrow, and side-by-side—they have continued to dry-harvest rather than sacrificing precious cranberry land for diking or irrigation ponds required by wet-harvesting methods. (Above, SBeebe; below, PCCRF.)

Ilwaco Cranberry Company, Incorporated

The Ilwaco Cranberry Company was incorporated under the laws of the State of Washington, in December, 1914. With its principal place of business in Ilwaco, Washington.

The purposes for which this corporation is formed are: To ingage in, superintend, and carry on the business of developing, planting, growing and marketing cranberries and other fruits and agricultural products, and to can, preserve and otherwise prepare the same for market and in conducting and carrying on said business to buy, sell, option, contract for, and act as agents for other corporations and individuals, in the selling, buying, packing, canning, preserving and marketing of all such products,

2nd. To hold, own, purchase, lease or otherwise acquire, mortgage, sell, let, or otherwise dispose of lands, and to subdivide plat and improve the same and generally to carry on any other business which may seem to the corporation capable of being conveniently or properly carried on in connection with the above.

OFFICERS, DIRECTORS AND STOCKHOLDERS.

The officers, directors and stockholders of this company are all resident substantial business men of Ilwaco. Among whom are Dr. Lee W. Paul, Physician and Mayor of Ilwaco.

Robert M. Watson, Owner and Editor of the Ilwaco Tribune.

Ernest F. Saylor, President of North Shore Light & Power Co.

R. S. Jennings, Accountant and Train Dispatcher O. W. R. & N. Co.

George Clark, on construction work of North Jetty at Fort Canby.

J. W. Howerton, Proprietor of The Sprague Hotel and Real Estate Dealer.

This company has for sale cranberry land in large or small tracts. Its purpose is to develop, plant and care for the bogs until they become full bearing; to gather and market the crops if required, for non-residents, who are now or may in future become owners or investors; to assist and encourage the homeseeker in every legitimate manner within their power, that the growing of cranberries may develop into one of the leading agricultural industries of the State.

Respectfully submitted,

J. W. HOWERTON,
Secretary Ilwaco Cranberry Co.

SYNDICATES AND PROMOTERS. Among Washington's cranberry historians, the 1910s are known as "the Decade of the Syndicates." Most of these companies (which sometimes called themselves "exchanges") were out-of-state operations, though a few were owned and managed locally on the Long Beach Peninsula. They speculated in the purchase and sale of marshland for cranberry farms. Within a very short period of time, they controlled several thousand acres of marshland, which they promoted as ideal for cranberry growing. "One of the Leading Horticultural Industries of the Northwestern States," proclaimed one brochure. "Its purpose is to develop, plant and care for the bogs until they become full-bearing," promised another. Although the "cranberry boom" these promoters created was responsible for dramatically increasing the number of farmers and acreage devoted to the industry, it was doomed to be short-lived. The pest problems with the imported vines continued, and by 1917, losses were so great that many growers again lost faith in the industry. Those who remained began searching for help. (AMcPhail.)

WASHINGTON STATE COLLEGE. In 1913, a year ahead of similar federal legislation, the Washington legislature authorized Washington State College to become headquarters for the Bureau of Farm Development, an arrangement still in effect under the state university system. With the motto "helping farmers to help themselves," the bureau provided for the appointment and maintenance of agricultural experts across the state. It was to this bureau that Henry S. Gane, president of the Peninsular Cranberry Company, wrote year after year asking for help for the Peninsula's cranberry growers. His correspondence began in 1917 as a follow-up to a meeting with Dr. E.O. Holland, president of the State College at Pullman. Year after year, his pleas were noted but not acted upon. (Above, WSULibraries; below, CWREC.)

D.J. "Jim" Crowley. Finally, in 1922, assistance arrived for discouraged cranberry farmers. D.J. "Jim" Crowley, with a BS degree in plant pathology, was sent to the Long Beach Peninsula by the State College of Washington. His assignment was to investigate the problems of the cranberry industry on the Washington coast. Recognizing that the climatic conditions on the coast differed from other cranberry sections of the country, Crowley began experimenting with methods that would improve cultivation and crop yields under coastal conditions. His greatest contribution to horticulture was in the area of frost injury prevention. His recommendations were based on what he was taught in physics concerning water releasing heat as it freezes. Growers were hard to convince, but eventually, the sprinkler systems he suggested became standard equipment for commercial cranberry growers and other fruit growers throughout the world. (PCCRF.)

THE RESEARCH STATION. Without a laboratory or a car during his first two years on the job, Crowley conducted his experiments in a shed on the property near where he was living. He made his inspection tours of the 400 acres of cranberry farmland by train, bicycle, or sometimes, on foot, walking from Ilwaco to Oysterville, a round-trip of 40 miles. In 1923, twelve acres of uncleared and undeveloped land along Pioneer Road in Long Beach was rented from Pacific County for an indefinite period at $1 per year. The following year, the state legislature appropriated $9,000 to establish a "Cranberry Investigations Laboratory on the Coast," and a two-story laboratory and shop building was erected on the property. Crowley used the top floor as his laboratory and the first floor for equipment storage and a toolshed. That year also, land adjacent to the Research Station was cleared and, by 1930, four acres were planted with 20 varieties of cranberries. (Both, PCCRF.)

Two

Planting and Growing

Wetland Characteristics. Early cranberry growers were primarily interested in the peaty soil of the marshlands as potential sites that would support bogs of domestic cranberries. Acres of such land could be found on the Long Beach Peninsula, in the vicinity of Grayland, and north to Copalis Crossing. Today, those marshy areas, classified as "wetlands," are better understood for their environmental potential and are highly regulated. (PCCRF.)

STEWARDS OF LAND AND WATER. Like all other wetlands, the cranberry wetlands system plays an increasingly important role in water storage and conservation, in recharging aquifers and filtering out pollutants, in providing wildlife habitat, and in the preservation of open space. Cranberry growers are, by nature and necessity, masters of water management and careful caretakers of their watery wetland world. In addition to their cultivated bogs, cranberry farmers typically own an acre of wetlands for every cultivated acre of bogs. Cranberry bogs and their irrigation systems come under close public scrutiny. Federal, state, and county agencies who govern wetlands policy include the US Army Corps of Engineers, the Environmental Protection Agency, the Washington State Department of Ecology, the Washington State Department of Fish and Wildlife, the Washington State Department of Agriculture, the Washington State Department of Natural Resources, and Pacific and Grays Harbor Counties Departments of Community Development. (Both, AMcPhail.)

CLEARING THE LAND. One of the first requirements of a cranberry farmer is a storage facility for tools and equipment. Whether the marshland designated to become a cranberry bog is vegetated or non-vegetated, considerable work is required for bog preparation, and in the coastal climate, adequate storage for tools and supplies is a necessary early investment. In the late 19th century, when pioneers were experimenting with the first domestic bogs on Washington's coast, all the work was done by hand. Grubbing out an area dense with trees, shrubs, and other plants was difficult, time-consuming work, as was preparing the bog, planting, weeding, and harvesting. As mechanized tools were developed, farmers began to move away from the labor-intensive handwork. Gradually, things improved. (Above, PCCRF; below, BCoppin.)

DITCHING THE BOGS. Marshlands are nature's sponges; they store and purify water and help to maintain the water table. Most importantly to cranberry growers, vines thrive on the special combination of soils and water found in marshes. However, in order to regulate the quantity of water on the bogs, it is necessary to install ditches as part of bog preparation. In the wet northwest climate, ditches are necessary for drainage and, in the case of bogs that are wet-harvested, ditches can also be used for flooding. In order to keep the network of ditches free flowing, ditch cleaning is usually done in the spring and fall by hand or with a mini-excavator. (Above, CPHM; below, CGM.)

NORTH BAY BOGS. In 1937, the Hoquiam Chamber of Commerce committed itself to bringing the cranberry industry of North Bay into "full blossom." The plan called for developing the entire area as one unit in order to make optimum use of the land. Bogs were to be five acres or less per family, giving 240 families a "credible livelihood." Since there was little area for the installation of dikes, the configuration committed the farmers of these bogs to dry-harvest methods. The plan specified inclusion of a road that would make all bogs accessible. It took careful planning on the part of the Hoquiam Chamber of Commerce and the engineering staff of Grays Harbor County, approval of the plan by extension agent D.J. Crowley, and the cooperation of 18 bog owners to create the five-mile-long Burrows Road, shown below. (Both, S/GThompson.)

SANDING THE BOGS. Sand is such an important part of bog installation and maintenance that special legislation was created in Washington to allow Long Beach Peninsula growers to take sand from the nearby Pacific Ocean beach. Sand is needed for the initial preparation of bogs. Even though sand is plentiful along the coast, before mechanization, getting it from beach to bog was a laborious process. In the 1920s, early Oysterville farmer Alexander Holman hand-carried sand the mile from beach to his Oysterville bog in five-gallon buckets. He balanced the buckets from either end of a yoke-like arrangement hanging from his shoulders. His pride in his prizewinning bog (a record 208 barrels per acre in 1918) was only surpassed by his satisfaction at having voted for the first time in 1864, casting his ballot for Abraham Lincoln. (Above, JBogh; below, CJacobs.)

METICULOUS HAND-PREPARATION. Using only hand tools, bog preparation was slow work in the early 20th century. Typically, growers prepared and planted their bogs one section at a time, completing the first section before beginning the next. That way, their cranberry crop could begin paying for itself in a timelier manner. Depending upon the final size of the operation, the first sections might be ready for harvest as the last sections were being completed. Since it usually takes four to five years for a harvestable crop, this method helped defray some of the expenses as the farmer progressed. Once a bog is established, it has a long life. Some bogs on the Peninsula have seen 80 harvests and are still producing well. (Above, PCCRF; below, JBogh.)

THE SCORING METHOD. Before mechanization, bogs were hand-built and then hand-planted. The fully prepared bog was scored using a long, handmade wooden rake with wooden tines set about eight inches apart to line the sand. A steady hand, a good eye and, perhaps, a helper were needed to assure straight lines. Lines were scored first in one direction and then crossways, making a neat grid to guide the planting. Cuttings were carefully planted at each intersection, and the vines were irrigated two or three times per day for several weeks until the cuttings produced roots and new vine growth began. The vines grow slowly at first, but once a good root system is established they grow more quickly and fill the bed with a solid mat of vines. It takes about four years to produce a good crop of fruit from a new bed and up to six years before a new bed is in full production. (Above, AMcPhail; below, CPHM.)

TENDER LOVING CARE. With one- and two-year plantings, the emphasis is on hand-weeding to make sure that the plants stay weed free as the vines grow and fill in. Through the growing season, new plantings are fertilized weekly with nitrogen. Usually, a bog can be harvested the third or fourth year and produce a full crop by the fifth year. As a rule of thumb, every four square feet of vines will produce about a pound of fruit, assuming good colonization of the ground, an upright density of about 400–600 uprights for every two square feet, depending on the variety, and good pollination of the flowers. (Above, CGM; below, PCCRF.)

CRANGUYMA FARMS. Guy C. Myers was an investment banker with headquarters on New York's Wall Street and at the Olympic Hotel in Seattle. In 1940, he bought acreage in the center of the Peninsula, ultimately accumulating 1,000 acres, 93 of which he put into cranberry bogs. He named his farm Cranguyma, the "Cran" for cranberries, the "guy" for his own name, and the last three letters, "yma," a reverse spelling of his wife's name, Amy. During the next several decades, he continued developing the farm to include blueberries, raspberries, and a nursery in which rhododendrons and azaleas were raised. A processing plant at the south end of the property produced a variety of cranberry products including sauces, juices, cranberry-flavored sherbet, and more. Today, Cranguyma Farms is owned and operated by fourth- and fifth-generation family members. (Both, CGM.)

EARLY DISC PLANTERS. The first mechanical planters were single- and multiple-disc models that were pushed by hand over the bogs. With this method, vines were spread randomly on the sand at the rate of about two tons per acre, after which the farmer went over them with the disc planter that pushed them into the sand with one or more straight dull disks. While an improvement time-wise over hand planting each vine fragment, pushing the disc planters was difficult work no matter whether it was a single- or multiple-disc arrangement. Farmers often shared equipment and helped one another out with the more difficult tasks. (Both, PCCRF.)

LARGE DISC PLANTER. The large disc planter pictured here has eased the labor of planting manyfold. This planter is pulled by a tractor and, like the smaller planters, is moved across the soil in both directions. Generally, vines are baled for ease of handling and then spaced out across the bog to make sure there are enough for coverage. The vines are then scattered by hand before the planter proceeds. The goal is for the vines to go about three inches deep into the soil. When this large planter is used, hand planters are needed to plant the edges and corners that can't be reached with the big machine. In Long Beach, growers share the use of this planter, each grower pulling it behind his own tractor and then turning it over to the next grower in line. (Both, AMcPhail.)

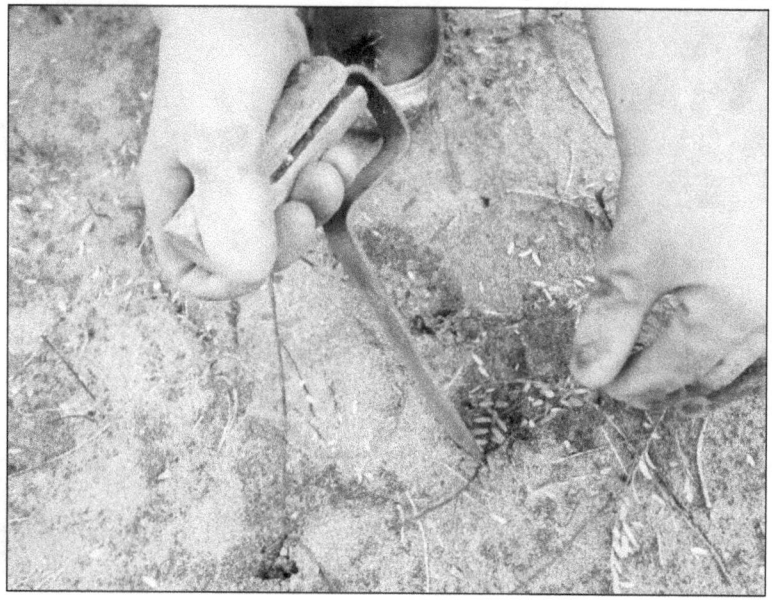

THE PLUG PLANTING METHOD. Rooted cuttings or plugs are more tedious to plant and are somewhat more expensive than scattering vines. However, they grow faster. Stolons (the uprights on which the fruit will grow) are cut and rooted, or "plugged," into large trays and are grown commercially in greenhouses. They are delivered to farmers when they have rooted and are ready to plant. Because they are already rooted, they grow faster than vines simply mowed from the fields. The operator has a tray of plugs that he drops into a chute that sets the plugs into the ground about every six inches. Commercially grown plugs have been available for about 10 years and are easily planted with the plug planter, which, as its name implies, is designed specifically for that purpose. (Both, AMcPhail.)

BUCKET BY BUCKET. Spreading clean sand on established cranberry bogs is a practice that dates back to the commercial birth of the industry when the first farmer discovered that sand blowing onto his cranberry vines stimulated their growth. He found that as the vines were pressed earthward into the soil from the weight of the settling sand, they gradually took root and soon formed new uprights on which berries grew. Spreading sand over the vines quickly became a regular practice. Traditionally, the sand was carried bucket by bucket every three or four years but, as mechanical improvements to other aspects of the industry took hold, farmers began to look for methods of sand delivery beyond the tried-and-true bucket-and-shovel method. However, for bogs that just need a little touch-up sanding, the bucket method is still used. (Both, AMcPhail.)

SLURRY SPRAYING. Spraying a slurry of water and sand has become a popular method of sand distribution. Depending upon the capacity of hose and nozzle, the sand-to-water ratio can provide good coverage in a relatively short amount of time. Sanding the bogs every few years provides growers with many benefits including insect and disease suppression, improved drainage, better root growth, and some frost protection, in turn leading to lower pesticide, fertilizer, and water requirements. In addition, sanding covers the leaf litter (trash) layer on the floor of the bog, which stimulates the decomposition of organic matter. It also helps suppress fruit-rot fungus and limits the habitat of cranberry girdler larvae, the scourge of dry-harvest bogs. (PCCRF.)

THE IMPORTANCE OF WATER. Water is the single most important resource needed in growing cranberries. As a general rule, each acre of cranberries will use four to five acre-feet of water a year to meet production and harvesting needs and both dry- and wet-harvested bogs use similar amounts of water for frost protection and irrigation. Most farmers have pumps that draw from surface ponds. Because bog sizes are limited in the Grayland and North Bay areas, the ponds are small and farmers often put down 10- or 12-foot well-points in order to recharge their ponds. Along the coast, where the water table is only 12 to 24 inches below the soil surface and where ponds are larger and deeper, well-points are not required. When there are extended frost nights or several hot days, the ponds can get quite low and may take some time to recharge. Wet-harvest bogs are constructed with elevated sides or dikes around their perimeters. These act as water barriers so the bogs can be flooded for harvest and, occasionally, for pest control. (Above, PKrebs; below, AMcPhail.)

IRRIGATION METHODS. Cranberries need about one-half inch of water a week, and growers bring water into the bogs in two ways: through flooding and through sprinklers. Wet-harvesters use about one acre-foot more water than dry-harvesters, but most of the water used during the harvest process is reused by transporting it through a system of culverts to the next bog in line. While all farmers, wet or dry, use sprinklers for irrigation and frost control, the systems can be simple or sophisticated, depending upon the needs of the farm. Sprinklers are generally spaced about 40 feet apart and drain lines are placed strategically, about 20 feet apart, along the sprinkler lines. When it rains, growers estimate amounts and, when necessary, supplement with sprinklers. In warm, dry weather, sprinklers run most days. Best management practices recommend irrigating in early morning to minimize the overall time plants are wet as well as to minimize loss from evaporation, run-off, and drift, which can amount to 30 percent of water that comes out of the nozzle. Irrigation is a major expense for cranberry growers. (Both, AMcPhail.)

Bog Rails. Before wet-harvests became possible, tracks were an essential part of bog infrastructure. No matter the size of the farm, rails were used to transport men, equipment, hand tools, buckets, picking boxes, field boxes, weeds, berries, vines, cleared stumps, brush, and anything that was needed to be moved to or from the bog. During its height, Cranguyma Farms, with its 93 acres of planted bogs, had five miles of standard-gauge railroad tracks. It was along the rails that harvested berries were hauled from the fields to warehouse for cleaning and storing. At times, when foot traffic was taking precedence, planks were used to cover a rail for easier walking. Above, during a field day at the Research Station, a single line of planks was apparently deemed sufficient for pedestrian use between the rails. In areas such as Grayland and North Bay, where there is no space for roadways adjacent to most bogs, farmers still depend upon their rail systems. (Both, PCCRF.)

From Simple to Complex. A simple rail line was often one of the first installations on a new bog. Not only was it helpful in the initial stages of creating a bog but it also served to support the farmer's work at every step. Often, the cars that ran along the tracks were homemade affairs, equipped with portable boxes or barrels for various purposes. Above, the workers appear to have used the tracks simply as a footpath or, perhaps, with a rail cart, just out of sight. Below, the farmer is using the rails to deliver a chemical or a fertilizer to his crop. The barrel aboard the cart serves as a "nurse tank" for mixing in and, then, spraying out. Properly managing the amount of fertilizer used on cranberry bogs can be both environmentally and financially beneficial. (Both, CPHM.)

FARM BUILDINGS. Like all farming operations, growing cranberries requires a variety of buildings, some of which are dedicated to specific purposes. Warehouses are often all-purpose buildings, serving as storage areas for equipment and tools. These days, regulations require a separate building for any chemicals and must meet rigid specifications. In the picture above, the grower's house is much smaller than the warehouse, typical of many farm situations. In the early days when all berries were dry-harvested, many drying sheds like the one pictured below were a necessity. They provided space out of the weather for the drying trays on which newly harvested berries were spread until delivery to market. Today, coastal Washington berries destined for the fresh market are sent directly from the bog to a processing and packaging site in Selah, Washington. (Both, PCCRF.)

FROM MODEST TO GRANDIOSE. Long before the adage "form follows function" became the watchword of 20th-century architecture, farmers instinctively knew how to design for their needs. The small pump house above is not only adequate for sheltering the farmer's pump but also presents an idyllic picture reflected on a sunny day. The Cranguyma greenhouse, on the other hand, is regal in its proportions and is perfectly suited to the needs of the grower. With an eye toward developing new varieties and expanding his operation, Guy C. Meyers built his greenhouse capable of holding 20,000 one-gallon plants. In 1946, he hired Rutgers University plant-breeder Dr. J. Harold Clarke to serve as manager of Cranguyma Farms. Clarke, already well-known for his specialized knowledge on cranberry, blueberry, and other small fruit production, would later became a world-renowned figure in rhododendron horticulture. (Above, C/JWarness; below, CGM.)

WEATHER STATION.

WEATHER STATION. In 1953, the Long Beach Research Station became one of the nation's volunteer sites for monitoring weather. That year, the US Weather Bureau determined that there should be one weather station every 25 miles for tracking nationwide rainfall. By 1990, the network included 10,000 sites. Volunteers have collected and submitted data from the stations over the years. This has provided an important archive for the nation's weather history and serves as a resource for research on global climate change. The weather station pictured here is still in use, although a newer one with added technology and networking ability was installed in 2005. Some growers also have their own weather monitoring systems for keeping track of temperatures and automatically activating sprinkling systems when needed. (Both, PCCRF.)

Three
TENDING AND HARVESTING

KILLER SPRING FROST. In an annual report for the 1925 *Washington State College Agricultural Bulletin*, extension agent Crowley wrote, "Cranberry bogs are especially subject to frost injury. Even as late as the end of June, temperatures which cause blossom injury have been noted. Most of the frost damage occurs in the late spring and methods for preventing this type of injury will receive much attention during the coming year." (CGM.)

WIND MACHINE. Since cranberry bogs are usually the lowest feature in the landscape, they typically can be up to 10 degrees cooler than the surrounding uplands. According to published guidelines, the lowest safe temperature that cranberries can withstand is 23 degrees Fahrenheit, and since spring nights often fall within the danger range, growers face an ongoing challenge. Although early growers had a clear knowledge of the problem, finding a workable remedy was another matter. One frost control method that was tried was the wind machine. According to proponents' theory, frost could be prevented by keeping the air moving over the bog so that the cold air would not settle onto the fragile cranberry plants. The method was only minimally effective. (Above, CWREC; below, PCCRF.)

SMUDGE POTS. A once familiar sight in orange groves, smudge pots are oil-burning devices placed between fruit trees for the prevention of frost. The burning oil creates some heat but, more importantly, a large amount of smoke forms a "blanket" that blocks infrared light, thereby preventing radiative cooling that would otherwise cause or worsen frost. Positive results found by orchardists prompted cranberry growers to experiment with the method, but they found that a great deal more heat would be required to warm air clear to the ground. The report concluded smudging might be of value in light frosts but when temperatures went down to 24 degrees Fahrenheit or below, it was ineffective. In an *Oregon State College Bulletin*, it was also noted that the pots cost 42 cents each and the diesel oil used for smudging, about 6¢ a gallon. About 70 to 125 pots per acre would be needed. (Both, PCCRF.)

Frost Control. In 1954, the year he retired, Crowley was continuing to espouse the positive reasons for sprinkling: "Growing cranberries without frost protection is very much of a gamble since an entire crop may be destroyed by one frosty night. Control of frost by sprinkling is effected in several ways. The temperature of the water used for sprinkling is generally about 45 degrees Fahrenheit. This water, when sprinkled on the bog, releases heat as it cools to the freezing point. The chief protection, however, comes from the releasing of the latent heat of fusion." Though he had done his first frost control in Washington bogs by sprinkling in 1925 and 1926, Crowley reported that 30 years later, only about three-fourths of the acreage was protected by sprinklers at the time of his retirement. (Above, CGM; below, PCCRF.)

```
PHOTO NO. W-2700-6    DATE 4/9/62
SCD  Grays Harbor        STATE  Wn.
LOCATION  NE Grayland
OPERATOR  Joe Boss
PHOTOGRAPHER  J. Ferguson
LEGEND State fully what, how, who and why
  Time for another fitting (tee). Just use
  a tape to measure distance and a hand saw
  to cut the pipe; swab on some thinner and
  cement and shove together and you're
  ready to move on. (Plastic) systems are
  installed quickly. Wayne Karjala helping
  his uncle, Wm. Roiko.



SCS-16  2-57                              USDA-SCS
```

THE RESEARCH LAB. The cranberry research lab is set up to do the following: insect and disease identification, fruit quality measurements (soluble solids, color, firmness), quantification of fruit rot, seed set, upright leaf area, and fruit-set, soil pH and texture measurements, raising of parasitic insects, testing of biological controls, bioassays for soil health, and herbicide activity. Tests are done continuously on the bogs around the research station and may be requested by growers as well. In addition to these activities, the extension agent keeps growers abreast of new information through workshops, seminars, conferences, one-on-one visits, on-farm demonstrations, field trips, and tours. He also keeps abreast of current research information, develops computer applications, develops and implements evaluation plans, recruits and utilizes volunteers, and reports results to the growers, the administration, and the public. (Both, T/MFC.)

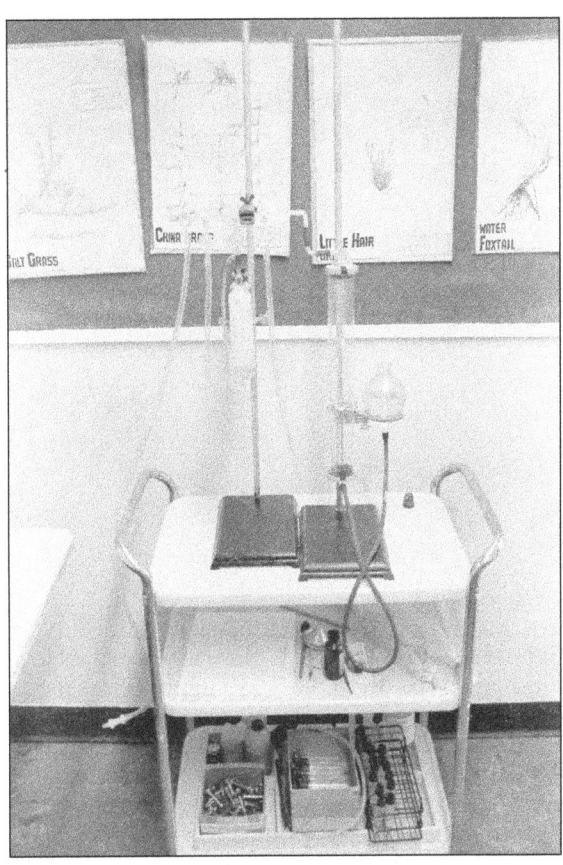

SIGNIFICANT PESTS. Several insects are major problems to cranberries. Two in particular, the black vine weevil and the girdler, are a significant problem in Grayland and North Bay as they thrive in dry-harvested bogs. Larval damage to cranberry roots by the black vine weevil (above) reduces the root system's ability to move water from the soil to the foliage. Damaged vines do not recover. The girdler attacks cranberry vines at or below the soil surface. The larvae remove the bark and conductive tissues of the stems, thereby "girdling" them and cutting off movement of water and nutrients in the plant. Flooding bogs is an effective control measure for both of these pests but not always possible in dry-harvest bogs. Insecticides are also effective. (Both, CWREC.)

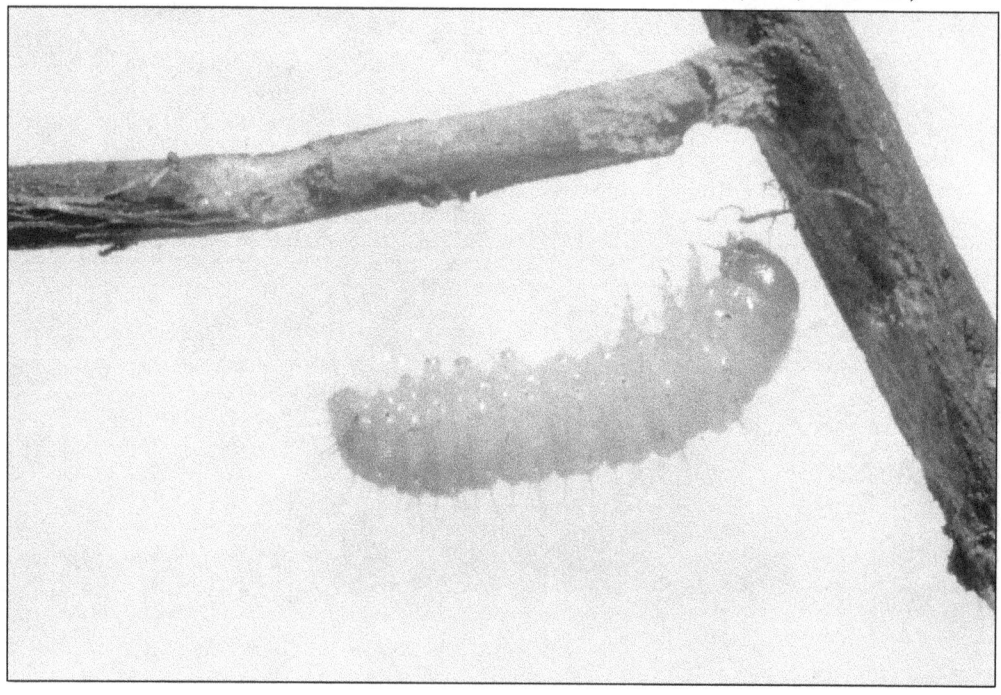

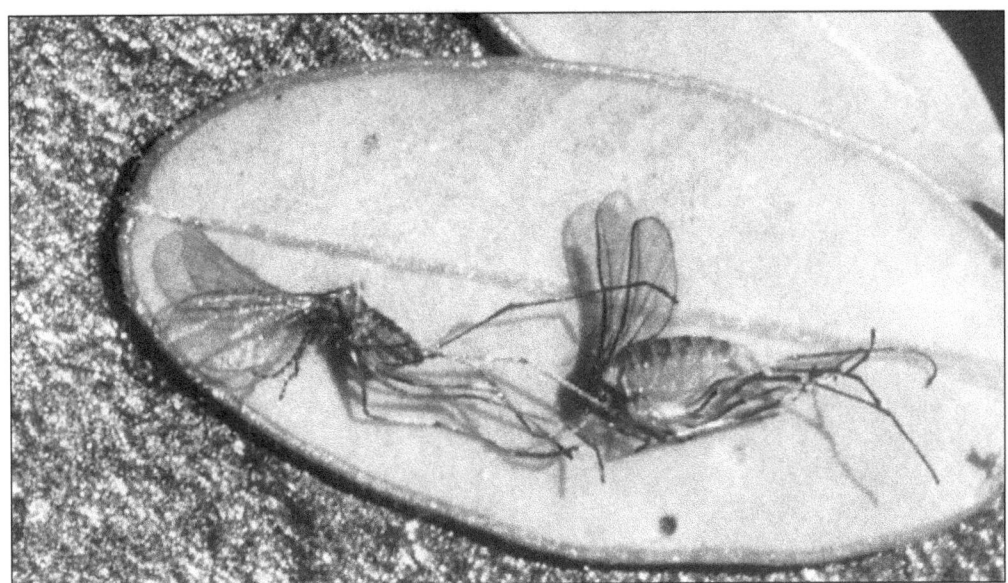

CONTINUOUS VIGILANCE. Cranberry tip worm (above) first appeared in Canada in 1998 and is now found throughout all cranberry-growing regions. One-tenth the size of a mosquito, it prefers the most vigorous uprights, reducing the terminal buds that are most able to convert to blossoms and fruit. To monitor for tip worm eggs and young larvae, cranberry uprights need to be examined under a dissecting microscope. Controlling tip worm is a challenge, since generations overlap in the field. Another significant pest is the black-headed fireworm, which overwinters in the egg stage on undersides of cranberry leaves. Larvae are a quarter-inch long at maturity and feed on new tip growth and runner ends, damaging berries and fruit buds for the next year's crop. Both the tip worm and the black-headed fireworm can be controlled with insecticides. (Both, CWREC.)

PEST CONTROL. Two of the most significant cranberry pests on the coast—the black vine weevil and the girdler—can be drowned by flooding the bog, as seen above; this is a possibility, unfortunately, that is not open to most growers in Grayland and North Bay. The most obvious sign of black vine weevil activity is adult feeding on foliage, which appears as notching along the leaf margin. Though the leaf damage, in itself, is of little economic importance, it alerts farmers to the much larger problem, which is damage to roots. Spraying can be effective if flooding is not possible. Timing is important with any of these control methods in order to get the best results. Prior to spraying, farmers may go out with sweep nets at night to catch weevils in order to see how many are around. (Both, CWREC.)

PROTECTING THE CROP. In 1917, in one of their early requests for help, the Washington coast cranberry growers wrote to the US Department of Agriculture: "The worms at the present time are beyond our control and, unless we have the best expert advice and help, our bogs are doomed." It would be another five years before the first extension agent was assigned to the area, but from the first, D.J. Crowley worked continuously on ways to control and reduce the number of pests destroying the cranberry crop. As well as working on new products for the job, Crowley developed a program reducing the number of times spraying was required. Testing pesticide spray solutions for effectiveness against bugs, fungi, weeds, and other threats to the health of the cranberry crops continues to be a main focus of Long Beach extension agents. Field tests involve spraying different strength solutions on different areas, as well as leaving a "control area" that is not sprayed. Once the results are evaluated, recommendations can be made to the growers. (PCCRF.)

HELP FROM THE STATE. To date, four Washington State College (University since 1959) scientists have served as Long Beach extension agents and directors of research: D.J. Crowley (serving 1922–1954); Dr. Charles C. Doughty (1954–1965); Dr. Azmi Y. Shawa (1965–1989); and Dr. Kim Patten (1990–2019). Each contributed to the solution of ongoing problems and aided in the ever-increasing success of Washington's cranberry industry. Crowley (above) arrived in Long Beach determined to find a solution to the killer frost that crippled so many bogs each spring. His eventual solution would benefit Washington's cranberry farmers as well as fruit growers throughout the world. After Crowley's retirement in 1954, Doughty focused on controlling twig blight fungus, another scourge for many coastal growers. It was during this period that public sentiment toward pesticides made it necessary to redo work previously accomplished on herbicides. (Above, PCCRF; left, CWREC.)

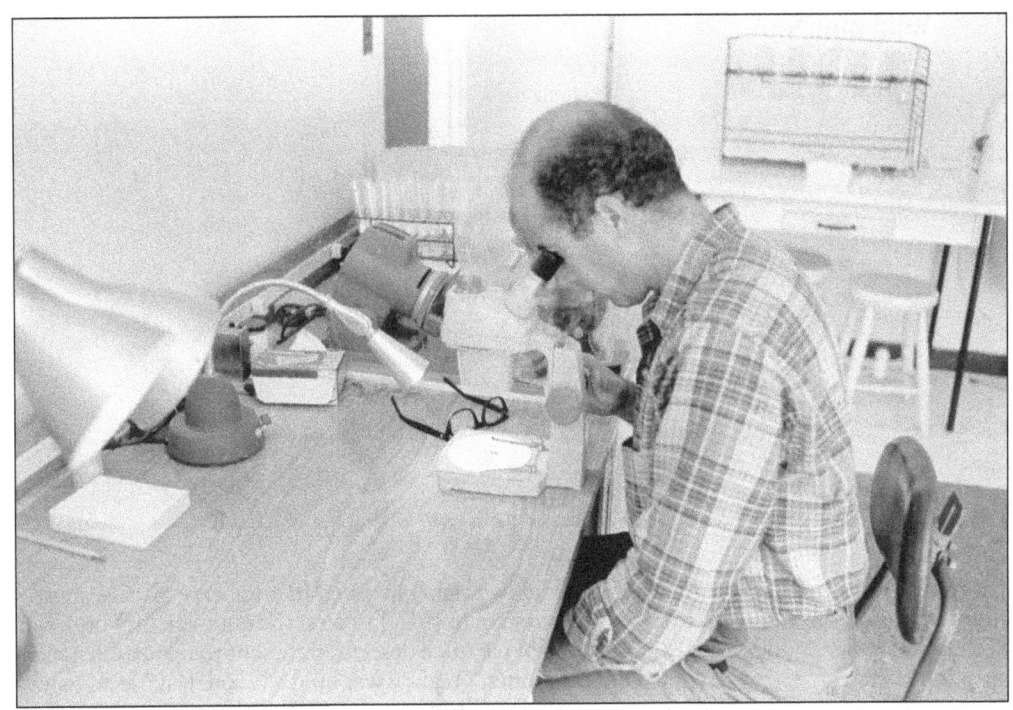

UNRELENTING RESEARCH AND TESTING. When Dr. Shawa (above) became horticulturist and area county extension agent for both Pacific and Grays Harbor Counties in 1965, he devoted his attention to pesticide and herbicide research in addition to cranberry fertility and color enhancement. By the time Dr. Patten arrived as associate horticulturist and extension agent in 1990, the station was well-established as a major contributor to the cranberry industry in Washington, Oregon, and beyond. Patten's major contribution has been in the area of pest control and in research on environmentally specific and safe uses of insecticides and herbicides. In addition, his continuing field studies concerning the most suitable and reliable cranberry varieties have resulted in reciprocal relationships within the industry. (Above, CWREC; below, AMcPhail.)

TROUBLES BEYOND THE BOG. Ira Murakami (above right) invested in cranberry acreage shortly after arriving on the Peninsula from his native Japan in 1917. Just after the outbreak of World War II, the Murakamis, along with 122,000 others of Japanese descent, were sent to relocation camps outside of areas considered restricted military zones. "Yeah, it was kind of a bad deal," Murakami's nephew Jeff said during a 1974 interview for an oral history series. "We sold our cranberry farm during the war while we were in the evacuation camp. We were forced to sell it because the person that was taking care of it recommended that we sell it." Shown here in 1920 with Murakami are, from left to right, local extension agent D.J. Crowley, Grays Harbor agent Vey Valentine, and Dr. Steich Wakabayaski, New Jersey State agent. Below, workers weed Murakami's bog. (Above, PCCRF; below, CPHM.)

BEE SEASON. Birds, bats, bees, butterflies, beetles, and various small mammals all pollinate plants, but for cranberries on coastal Washington, bumblebees are the best native pollinators. Unfortunately, there are just not enough of them to pollinate large areas, so cranberry growers on Washington's coast contract with regional beekeepers. Two to four hives per acre are required for pollination and are distributed throughout the bogs about the time the plants begin to blossom in late May. The hives remain on bogs until after July 4, about six weeks altogether. (Both, AMcPhail.)

NOT QUITE RECIPROCAL. Each blossom must be touched by a bee (or other pollinator) in order to produce fruit. Since male and female structures of cranberry blossoms mature at different times within a given flower, cross pollination is needed. In general, that means even more bee activity is required. Unfortunately, bees get few returns for their efforts because cranberry blossoms do not produce much nectar. The coastal climate can also adversely affect the pollination process as bees stay in their hives during rainy weather. One further hurdle in the pollination process is the downward-facing form of the flower, which discourages self-pollinating. Bog-watchers know that if the pink color of the cranberry flower intensifies with age, it indicates poor pollination. (MCrowley.)

FOR OPTIMUM FERTILIZATION. One of the earliest mechanisms for distributing fertilizer on the bogs was the hand crank type shown above. The riding spreader was a big improvement. It could broadcast granular fertilizer for 25 feet, thus saving a good deal of time and effort. According to experienced growers, a successful fertilizing program is more an art than a science. Phosphate and potash go on in early May, before the bees arrive. Nitrogen is applied when berries are pea size, usually after, but sometimes during, bee season. Growers evaluate the bloom as an indicator of how much future nitrogen might be required on each bog. They then evaluate the berry set, putting more nitrogen on heavy berry set. If a heavy crop is not adequately fertilized, the berries will "drain the vines," and next year's crop will be poor. (Both, AMcPhail.)

ANOTHER FARM CHORE. The short "bee season" means that farmers must keep a careful watch on pollination activities while the bogs are in blossom. If the bees refuse to leave their hives because of rainy weather, they must be fed. If the weather is warm and the temperature in the hives increases, the bees are likely to swarm, requiring a call to an expert to come in and assist in the move to a new hive. In addition, even though bees on cranberry bogs do not produce much honey, some farmers find it advantageous to install electric fences around the hives to discourage bears from making brood (larvae) raids. All of these bee-tending activities require time and attention. (Both, AMcPhail.)

BOG VISITORS. Cranberry bogs attract many visitors. In early spring, when other food sources are not readily available, bears may feast on rotten berries left from the previous year's harvest, but in general, they do not care for cranberries, nor do they like the wide-open spaces of bogs. Generally, they are just passing through in search of food or shelter elsewhere. Deer seek out weeds in the bog and trample vines while grazing. They also eat foliage or berries in late summer and may bed down in the bogs, damaging vines. Anecdotal evidence indicates that a single deer may eat two to four barrels of cranberries over the course of a season. If the animals become a nuisance, farmers may seek ways of scaring them away. Elk can be an even more serious threat to vines and sprinkler lines. (Both, AMcPhail.)

ONGOING VIGILANCE. Cranberry growing, like other sorts of farming, requires a lot of observing and checking things out. Not a day goes by that the vigilant cranberry farmer does not check and double-check some aspect of bog condition. Problems with weeds, pests, and fungus must be dealt with in a timely manner. Sprinkling systems and pumps must be kept in good condition. Bog infrastructure such as roads, dikes, and tracks need to be kept in good repair. Often, consultations with other growers or with the extension agent or a visiting expert can assist with a perplexing problem or give insights into new methods. "Seasonal work" to the cranberry farmer means work at every season, year-round. (Both, CWREC.)

WEEDS! WEEDS! WEEDS! Cranberry bogs are the perfect medium for weeds. On new bogs, tussocks and willows are the main offenders and respond well to wiping. Horsetail, "around since the dinosaurs," responds to only one chemical. Even harder to control are buttercup, dandelion, lotus, yellow weed (or yellow loosestrife), and false lily of the valley. These can only be pulled by hand-weeding when very small. Extension agent Patten tested (and got registered) a product that is safe for the cranberries and eradicates a lot of these broadleaf plants. It can be applied through the sprinklers and acts by removing the chlorophyll from the weed. Getting products registered for cranberries is difficult because companies are reluctant to put money and time into testing for such a small industry. Growers rely on researchers to get effective, new products registered. (Above, CPHM; right, PCCRF.)

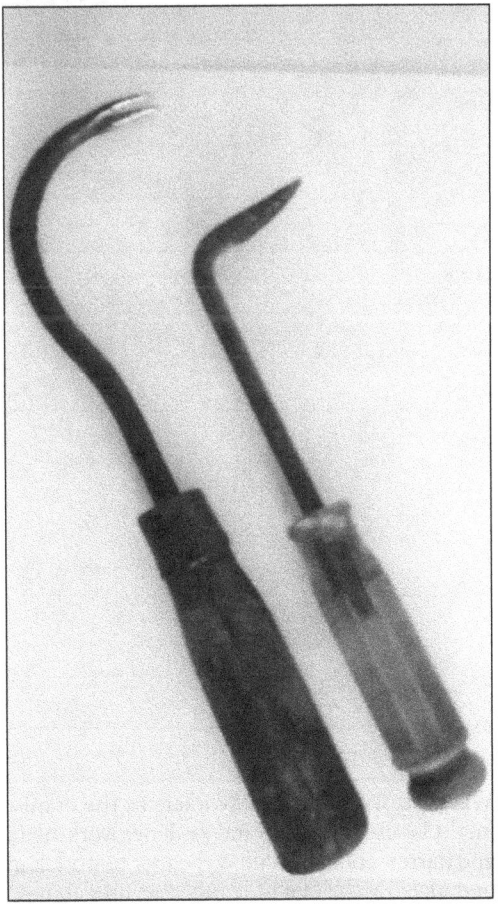

WEEDING WITH HOOKS. Workers in the cranberry bogs use weeding tools much like those that might be used by the home gardener working in a crowded garden bed. Implements must be long and narrow enough to allow the user to work it into the soil and down to the roots without crushing or damaging the mat of cranberry vines above. The tool also needs to disengage weed roots from

their grip around cranberry roots, again without damaging the cranberry plant. In former times, workers often fashioned their own tools—perhaps a reworked screwdriver or something made back in the day of the local blacksmith. No matter the tool, the work is tedious and time-consuming, best done with others for the companionship and friendly competition provided. (CPHM.)

AN ONGOING TASK. Keeping bogs weed-free requires relentless attention by farmers and their work crews. Stories of early, poorly maintained bogs that went irretrievably "back to nature" are not far from the minds of growers struggling to keep weeds under control. Assistance with weed management strategies is one of the primary responsibilities of the extension agent at the Cranberry Research Center and farmers are encouraged to take advantage of the strategies and techniques that are constantly being refined. The most frequent advice for new bog owners is "Don't let weeds get started." Finding a foolproof method of achieving that goal, however, is difficult indeed. (Above, CPHM; below, CWREC.)

WEED WIPING. The implement being used by the worker above is often called a "hockey stick" wiper. It is used to kill weeds that grow higher than the cranberries. Weed killer is put in the tube handle and seeps out onto a cotton fleece sleeve at the bottom. It is then wiped onto the weed, in this case, a tussock or sedge that thrives in wetland areas like cranberry bogs. Wiping can be done while weeds are actively growing. The bog shown below is a two-year-old planting that has been inundated with weeds. If the weeds are controlled, the vines will continue to fill in. Weeds need to be wiped before going to seed to eliminate the potential for additional weeds the following year. A strict pre-harvest interval must be adhered to after application of any pesticide. (Both, AMcPhail.)

BERRIES IN SUMMER. Growers often talk about fruit "sizing up" and "coloring up." A late bloom, a cold summer, or bad weather causing poor pollination can lead to poor fruit size and color. It is during the early fruit-set, too, that growers watch closely to determine if applications of fungicide need to be initiated. Berries are green at first and begin sizing and turning color in July and August. Due to the cool nights near the coast, the berries get much redder and darker than in any other region. West Coast fruit is prized for its dark color and is often mixed with berries from other areas to make juice darker. This dark color, however, is less desirable for making sweetened dried cranberries since the finished product looks too much like raisins. (Both, AMcPhail.)

HAND-PICKING BERRIES. From September when the first berries were ripe until the winter storms arrived in November, generations of farmers relied on large work crews to harvest their cranberry crops. Workers developed their own methods for picking the berries from the wiry vines. Protecting hands and fingers was necessary, yet wearing gloves interfered with the sense of touch necessary for careful work. Some workers wrapped their fingers in bandages and used a scooping motion to gather the berries, while others placed palms together and gently pulled the berries off the vines using both hands in pincer fashion; still others wore fingerless gloves. An ordinary laborer could fill six to twelve boxes a day while an expert could complete as many as twenty. (Above, CPHM; below, PCCRF.)

AN INTERNATIONAL WORK CREW. Keeping a year-round work crew for the hard, manual labor of raising cranberries has always been difficult on the sparsely populated Washington coast. Even today, growers regularly depend upon immigrant help. The men in the photograph above, probably taken in the 1930s, are identified as coming from Belgium, Bulgaria, England, America, Italy, Germany, Sicily, Sweden, and Scotland. During World War II, when manpower was scarce, there was a shortage of pickers everywhere. At that time, a full complement of pickers numbered 1,200 in Grayland, and they reported that they were 200 men short. They worked by hand using a variety of scoops, according to their personal preferences or what the growers had available. (Both, PCCRF.)

PRIDE IN THEIR WORK. That crews took pride in their work is evidenced by the photographic record, often showing groups such as this one posing with their equipment and, perhaps, a sample of the day's production. This photograph is notable for the young man at the top right who brought his string of ducks to be recorded for posterity. One wonders if it was a usual practice to bring a shotgun to work "just in case" and whether or not there was a protocol for dividing the day's bounty with the bog owner. Certainly, situated as they are right along the Pacific Flyway, the bogs of Washington's coast make ideal hunting grounds during the fall when migratory waterfowl are moving south and the harvest is in full swing. (PCCRF.)

LABOR-INTENSIVE. Gradually, the use of wooden scoops replaced hand-picking. At first, scoops were handmade wooden rake-like implements but gradually became available commercially in a wide variety ranging from small scoops for women and children to long-handled scoops for tall men. Some were double-handled for two-handed scooping and some were made of tin or other metal. Utilizing scoops rather than picking by hand was easier on the workers and made the harvest go more quickly. Even so, it was difficult, labor-intensive work and farmers were sometimes hard-pressed to find a crew. As the work progressed, pickers added their full hampers to the drying trays lined up beside the bog. These would be collected at the end of each workday and placed into the farmer's warehouse or drying shed to await delivery to the processing plant. (Above, PCCRF; below, FM.)

THE BOUNCE TEST. The best cranberries bounce. Presumably, it was a one-legged, mid-19th-century schoolteacher and cranberry grower, John "Peg Leg John" Webb, who discovered that only healthy, ripe cranberries bounce. According to the story, Webb was unable to carry his barrels of cranberries down from the storage attic, so he simply tipped the barrels and let the berries bounce down the stairs. He discovered that only the best cranberries made it to the bottom. He had inadvertently invented the "bounce board," still used today in various forms (such as the cranberry separator shown here) to determine cranberry quality. At modern processing plants, cranberries are still given the bounce test and unripe or damaged berries (such as the victims of cranberry fruitworm, seen below) are culled out immediately. Because the best berries bounce, cranberries are sometimes called "bounce berries." (Both, PCCRF.)

BOXES AND BARRELS. Almost since the beginning, West Coast growers have used boxes of various sizes for cranberry storage, rather than the traditional barrels used by growers in the eastern states. Even so, barrels remain the time-honored unit of measurement for cranberries, each barrel the equivalent of one hundred pounds. Growers speak of harvest yields in terms of barrels, though burlap sacks and boxes have long been in use by farmers who dry-harvest. Though wet-harvested berries are delivered straight into totes or trucks whose capacities are spoken of in terms of pounds or tons, *barrel* is the standard word for measuring cranberries everywhere, whether fresh, processed, dry-harvested, wet-harvested, or organic. Shown above (from left to right) are a picking measure, field box, shipping box, and storage tray. (Both, PCCRF.)

MECHANICAL PICKERS. Today, mechanical pickers have taken the place of hand-picking on dry-harvested bogs. The Darlington, or "Eastern," style picker was introduced shortly after World War II and was soon followed by the "Western" (shown here) developed in Bandon, Oregon. Both pickers greatly speeded up the harvest, but liabilities included the strength needed in controlling the machines and the high percentage of lost fruit due to bruising and the damage to vines during harvesting. In previous generations, wooden boxes were the collection device of choice but today most growers utilize burlap. Once the bag is full, the operator stops the machine, removes the bag and sets it on the bog, places an empty gunnysack onto the machine, and continues harvesting. (Both, PCCRF.)

PICKER/PRUNER. Like the Eastern and Western pickers, the Furford picker/pruner combs through cranberry vines, with its 14 tines guiding the berries to conveyor paddles and lifting them into a waiting burlap bag. The machine is powered by a five-horsepower engine bolted on top. In addition to picking the cranberries, the Furford simultaneously prunes, its 14 knives snipping the cranberry vines and gathering them along with the berries. They all land in a gunnysack hooked on the back of the machine. Built one at a time and made to last, the Furfords are still used extensively in Grayland and North Bay and in other areas where dry-harvest methods are practiced. The Furford was invented by longtime Grayland resident Julius Furford, who passed away in 1999 at the age of 91, still active in his two-man factory. (Both, PCCRF.)

FURFORD CRANBERRY MUSEUM. Constructed in 1933 by members of Grayland's Finnish community, this building was originally a social hall. During World War II, the Grayland and Long Beach growers used it to dehydrate berries for shipment to servicemen overseas. After the war, in 1946, Ocean Spray bought the property, constructed a second building, and used both for dehydrating berries. The buildings were vacated in the 1950s when Ocean Spray moved to the cooperative's current plant in Markham, Washington. Julius Furford purchased the property in 1956 and eventually converted it into a museum to chronicle and preserve the history of cranberry production in the Grayland area and to document the development of his picker/pruner. Today, Furford's dream of a cranberry museum "for the people, especially kids" thrives under the watchful eyes of Chuck and Gwen Tjernberg. (Both, FM.)

PRUNING AND PROPAGATING. Over time, vine growth can sometimes become very dense or the vines may get "leggy" and may need pruning. On dry-harvested bogs, pruning can be simultaneous with the harvest, thanks to the Furford picker/pruner, but in wet-harvested bogs, pruning is typically done, whether by hand or with mechanical assistance, after harvest is over. Pruning bogs keeps the canopy open so that air can move through and light can penetrate helping the fruit to develop and color. Conventional farming wisdom regarding pruning: "If you have only a few prunings you didn't fertilize enough through the season. If too many vines come off, you have over-fertilized. It's a happy medium and you may need to adjust your fertilizer schedule for next year." (Both, AMcPhail.)

THE SUCTION PICKER. Another innovation to come along was the suction picker. In the early 1950s, many types of these contraptions, reminiscent of giant vacuum cleaners, began to be seen on dry-harvest bogs. There were various models with one to multiple hose styles. All of them sucked the berries directly off the vines. While an improvement over hand-picking, the suction pickers were awkward to manage and somewhat labor-intensive, as they required a man at each hose plus one or more workers to move the equipment around, empty out the berries, and move them into the shed. Clarence Warness of Grayland claims that he invented the first suction picker when, as a teenager, he took his mother's new Electrolux out to the bog to suck up the berries. (Both, PCCRF.)

FRESH BERRIES. Unlike wet-harvested berries, dry-harvested cranberries can, with prior approval, be used for the fresh market. A premium is paid for fresh fruit, but growers wishing to produce fresh fruit are required to declare their intent ahead of harvest. The berries must meet cleanliness and firmness tests for keeping quality and are checked throughout the season as well. Immediately after they are harvested, these berries are placed in special containers and sent in semitrucks to a plant to be cleaned and packaged. They are soon seen on grocery shelves, ready for holiday use. Fresh, raw cranberries can be kept in the refrigerator for three to four weeks or can be safely frozen for ten to twelve months. There is no need for thawing the frozen berries before they are used for sauces or baking. (Above, S/GThompson; left, PCCRF.)

WATER-HARVESTED BERRIES. Using the scoops in the water made the berries float up, raising the vines with them and making it easier to rake off the berries. From these water scoops evolved the idea of "knocking" or "beating" the fruit off the vines with a reel with horizontal bars. Unlike dry-harvested berries, however, water-harvested berries cannot be sold fresh because: once they have been in water, they may rot more easily; in the knocking process, some berries get bruised; the natural, protective coating on the cranberry is washed off in the cleaning process at the plant so they do not keep. There are water-harvesting operations in Wisconsin with million-dollar drying facilities that can now get the fruit in and dried quickly enough to sell fresh, a possibility still not available to Washington growers. (Both, CPHM.)

WATER REELS AND BEATERS. A water reel is a machine used during a wet-harvest to release berries from the clockwise-trained vines by sweeping through the bog and knocking them loose. When the water reel is attached to a riding machine, it is called a "beater" and is used in a similar fashion. The first step in wet-harvesting is to flood the bog to a level just above the vines. The beater goes through and knocks the berries loose and, because the berries each contain four hollow seed cells, they float to the surface. Once the berries are freed from the vines, the water is raised higher to allow them to float more easily. Workers then surround them with booms, gathering them to one side of the field where they are loaded in trucks or totes. (Both, PCCRF.)

CORRALLING THE BERRIES. During this colorful period of the wet-harvest, workers use wooden booms to surround the berries. The booms are hooked together to form a corral within which workers use large paddles to move the floating berries toward the berry elevator. Wind can help or hinder the process, depending upon its direction. Volunteers sometimes pitch in, as well, and provide a little comic relief when they step in a hole and their boots fill, or worse, they trip over a sprinkler and get completely dunked. "All part of the process!" say the farmers. When berries reach the elevator, they are lifted into the waiting trucks. Slats on the bottom of the elevator allow the leaves to wash back into the water. These leaves, or "trash," will later be removed and used for mulch. (Above, AMcPhail; right, LSacks.)

DELIVERY TRUCK LINEUP. Harvested berries from the Long Beach Peninsula and from northern Oregon are hauled to Ocean Spray's Long Beach Receiving Station. They arrive in dump trucks or in trailers loaded with totes. At the receiving station, berries are moved along a conveyor belt as leaves and weeds are removed with brushes and blowers. The berries are bounced to remove the bad ones and then washed. When the cleaning is complete, they are loaded into totes and taken by semitrucks to freezer plants to await processing. Eventually, they will become sauce or juice or one of the many other Ocean Spray processed cranberry foods. Except for fruit destined for the fresh market, berries from Grayland and North Bay also come to Long Beach for cleaning. They are then sent to Markham, Washington, to be made into sweetened dried cranberries or into concentrate and shipped to Henderson, Nevada, for bottling into juice. To be sold fresh, dry-harvested fruit must "qualify" ahead of harvest. Once picked, the fruit goes immediately to be cleaned and packaged. (AMcPhail.)

Four

Processing and Marketing

THE LINEUP. Harvest is a busy time on the bogs and at the Ocean Spray delivery site in Long Beach. Berries arrive at the receiving station on Sandridge Road by dump truck or tote load and, on busy days, must wait in line for their turn to be received. Once the delivery process has been completed, the trucks return to the bog for their next load. (AMcPhail.)

WEIGHING IN. On arrival at Ocean Spray, each loaded truck is weighed before lining up to dump its load of berries into the hopper at the beginning of the cleaning line. Because of the relative sizes of hopper and truck, it is sometimes necessary for the truck to dump only part of its load and then wait until enough fruit has run through the line to allow dumping the rest. Each empty truck is reweighed to determine the weight of the berries for that load. Carefully labeled samples are taken from each load for testing in the Ocean Spray lab for size, color, firmness, rot, and "trash." Incentives are paid if berries meet specifications for each of these qualities, and growers are charged if the percentage of trash or rot is too high. (Above, T/MFC; below, AMcPhail.)

ON THE LINE. Except for fruit destined for the fresh market, all Washington coast berries grown by Ocean Spray Cooperative farmers go first to the receiving station in Long Beach (above) to be weighed, cleaned, sorted, and evaluated before being sent elsewhere. From the hopper, they go onto a conveyor belt—"the line"—that has been modernized to eliminate old, labor-intensive techniques. Gone are the days when workers were sometimes called "Lucy" in affectionate memory of the famous *I Love Lucy* episode in which she and Ethel work frantically on the chocolate-wrapping line in a candy factory. Independent growers do their own sorting and cleaning, often in a privately owned warehouse, and must make their own arrangements for processing, marketing, and distribution. (Above, T/MFC; below, CPHM.)

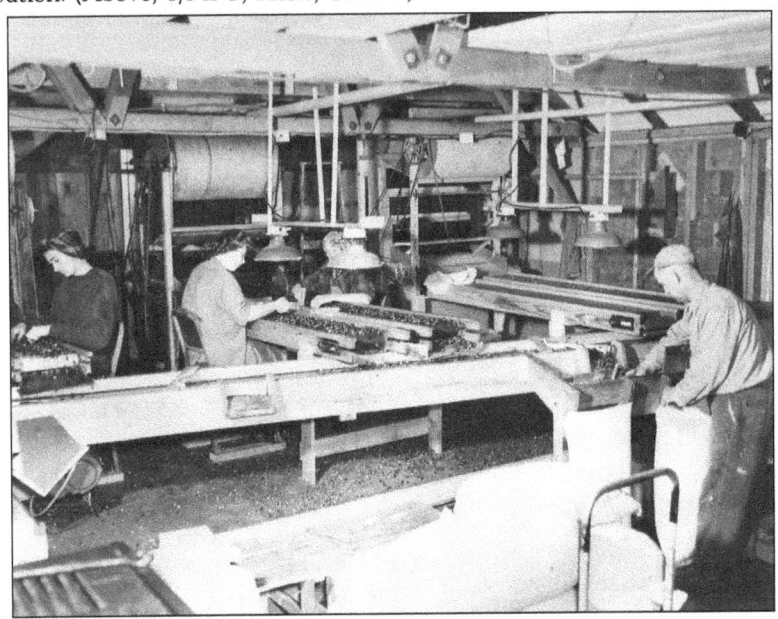

CRANBERRY RECEIVING STATIONS. The Long Beach receiving station, seen above, is one of three similar Ocean Spray facilities on the West Coast; the others are in Bandon, Oregon, and in Richmond, British Columbia. In addition, there are several processing plants, each specializing in one or more Ocean Spray products. Most berries from Washington coast's Ocean Spray growers are sent to freezers in Forest Grove, Oregon, until needed for making concentrate, which is the basis for juice and sauce. That process happens in Markham, Washington, in the plant seen below. Although the numbers occasionally change, as of 2017, there were 1,513 acres of Ocean Spray–affiliated cranberry bogs on the Washington coast (579 wet-picked and 924 dry-picked). In addition, 189 acres were grown by independent growers not affiliated with Ocean Spray. "Independents" must make other arrangements for their berries. (Above, AMcPhail; below, DBellamy.)

EARLY MARKETING METHODS. Before the Ocean Spray Cooperative was organized in 1930, most growers marketed their cranberries independently. Following the fall harvest, tempting displays proliferated along the most traveled roads near Long Beach, Grayland, and North Bay, and festivals and fairs were organized in order to capture the interest of area housewives. In addition to cranberry products, the fairs often included exhibits of agricultural livestock, flowers and confections, and exhibits by schoolchildren. Exhibitors came from Grays River, Deep River, Naselle, Oysterville, Nahcotta, Ocean Park, Seaview, Long Beach, Ilwaco, Chinook, and the places between and beyond. In promoting the 1932 fair, the Grange president said, "Everyone bring flowers for the flower exhibit and cranberries for the cranberry exhibit for such flowers and such cranberries are grown nowhere else upon earth but on this fine old Peninsula." (Above, KSnyder; below, CPHM.)

MARKETING WOES. During the first decades of the 20th century, growers on Washington's coast still felt isolated from the major market areas in California. Although access to rail transportation was increasing, most shipping in and out of the area was still by water. Shipping schedules were often adversely affected by the stormy fall weather, and there were no guarantees that fresh cranberries would arrive at their destination in marketable condition in time for the Thanksgiving and Christmas holidays. Some growers, like Guido Funke of Ilwaco, handled their own marketing and distribution. They found that investing in copper printing plates, packing labels, and other shipping paraphernalia was preferable to the uncertainties of independent sales companies. (Both, AMcPhail.)

¼ U.S. STD. CRAN. BBL.   PACKER No. _____

MIST-KIST
CRANBERRIES

PACKED FOR THE

NATIONAL CRANBERRY ASSOCIATION

A NATIONAL DISTRIBUTOR. In the belief that an established distributor with experience in the markets of the eastern United States, some Washington growers began doing business with the National Cranberry Association. Perhaps their repurposed labels were an indicator that their investment in the western market was tentative, yet they maintained a presence on Washington's coast throughout the 1940s. By 1946, they were packaging fresh cranberries in one-pound, cellophane bags, 20 bags to a box, a departure from their initial, rather limited interest in buying cranberries exclusively for canning. Their Ocean Spray label was a brand, not to be confused with the Ocean Spray Cooperative that is so familiar today. (Both, PCCRF.)

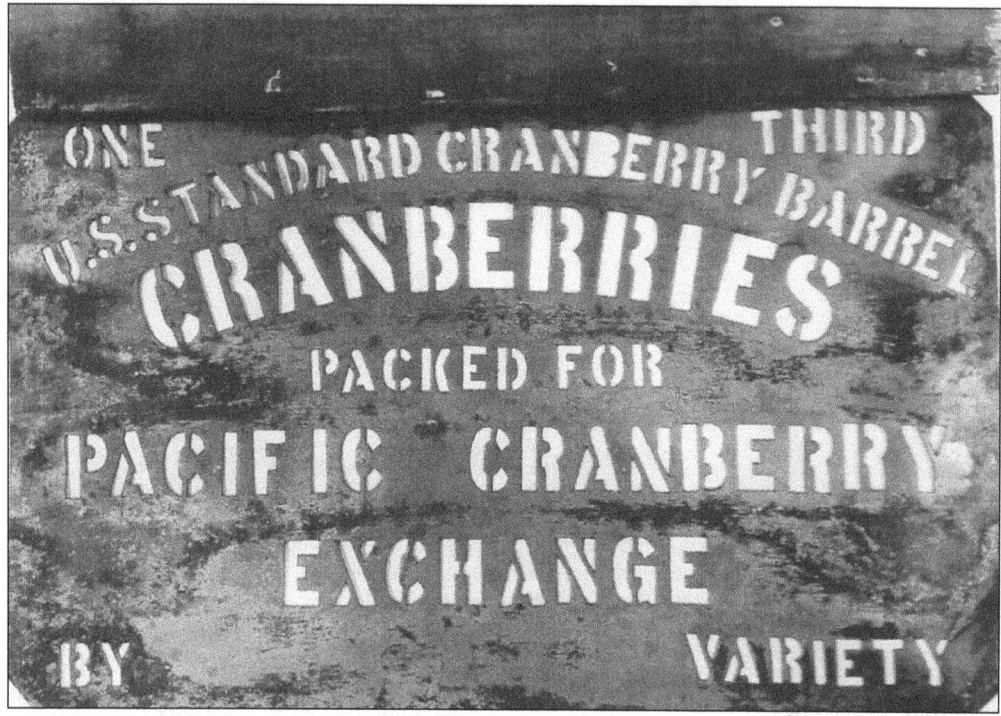

WESTERN COMPETITION. A nonprofit cooperative marketing association holding no stock and calling itself "Pacific Cranberry Exchange" was organized in 1917. Its purpose was to stabilize pricing and improve on quality of cranberries arriving at their target markets in San Francisco, Los Angeles, and San Diego. Seven years later, the association's first report noted that member growers "feel that the Eastern people have more or less taken advantage of the situation in the past, and therefore they are very glad to see a little competition started, and all of them give us a helping hand in opening the territory." In 1937, the exchange developed a new-style, lighter-weight veneer shipping box on which labels were painted by the factory. This proved to be a savings as well as a convenience. (Above, LCrowley; below, PCCRF.)

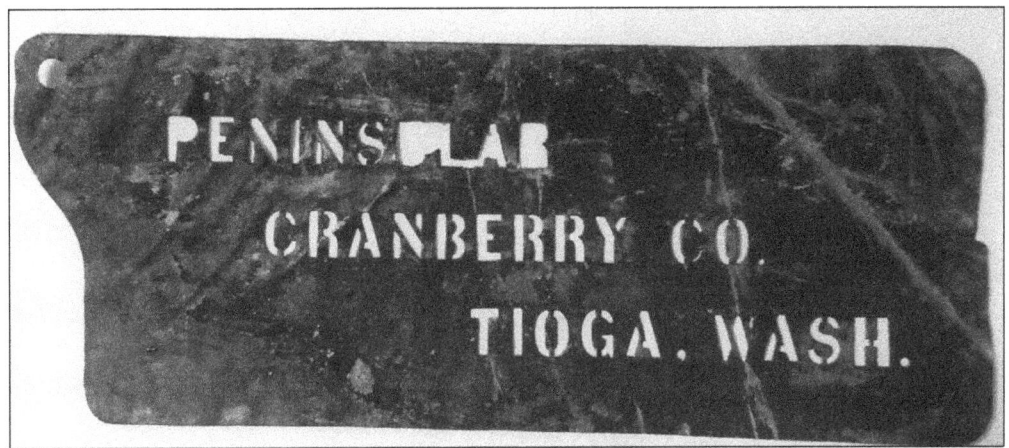

BEYOND THE EXCHANGES. Growers were constantly searching for markets outside the exchanges but it was a risky business. Stories linger about Henry S. Gane and his Peninsular label with its unusual play on the word *peninsula*. After much research, an investment of $60,000 in acreage, and a trip to New England to investigate cranberry varieties, he lost his first crop when his entire bog became infested with worms. Meanwhile, directly across the road on their independently owned, small three-acre plot called the "Gane Jr. Bog," his wife and children had a successful harvest with a different variety. Another enterprising Long Beach resident, Theo Jacobson, sold a special cranberry relish in his small grocery store. Though Jacobson was not a cranberry farmer himself, his sister owned a tiny bog nearby, and berries, recipe, and sauce may have well been a family endeavor. (Both, LCrowley.)

NORTH BEACH Cranberry RELISH

MADE FROM STRICTLY PURE WESTERN GROWN CRANBERRIES
A DELICIOUS SAUCE for ALL KINDS of MEATS

Made by THEO. JACOBSON & SON
Long Beach Washington

NET WEIGHT 12 OUNCES

CRANBERRIES IN TOWN CENTER. In July 1949, the Ocean Spray Cranberry Center opened in the Cranberry Canner's Building in Long Beach. The new shop featured many cranberry products as well as cranberry sundaes and milkshakes. The arrival of Ocean Spray in the area was big competition to other cranberry marketers and, especially, to Cranguyma Farms with their particular array of cranberry products including ice cream and cranberry toppings. Ocean Spray also offered a cannery, freezer, and storage space for growers, eliminating their need to store berries in their own packing sheds. However, it would be another dozen years before Washington growers had joined the Ocean Spray Cooperative in sufficient numbers to warrant the construction of the Ocean Spray Receiving Station in Long Beach. (Above, PCCRF; left, CGM.)

GIFT IDEAS. By the mid-1940s, Ocean Spray was advertising in magazines such as *Better Homes and Gardens* and *Woman's Day*, offering promotions such as the tray and spoon, seen above, from the Wm. Rogers Company for three sauce labels and one dollar. A recent survey showed that more than 94 percent of Thanksgiving dinners include cranberry sauce. The jellied variety (the log) is most preferred by consumers, totaling 75 percent of overall cranberry sauce sales. Cranguyma Farms on the Long Beach Peninsula became a national leader in the development of its own cranberry product line including variety gift boxes like the one pictured below. Cranguyma, at one time, employed more than 50 workers and included a cannery, a packaging plant, a nursery, and gardens, in addition to its cranberry bogs. (Above, AMcPhail; below, CGM.)

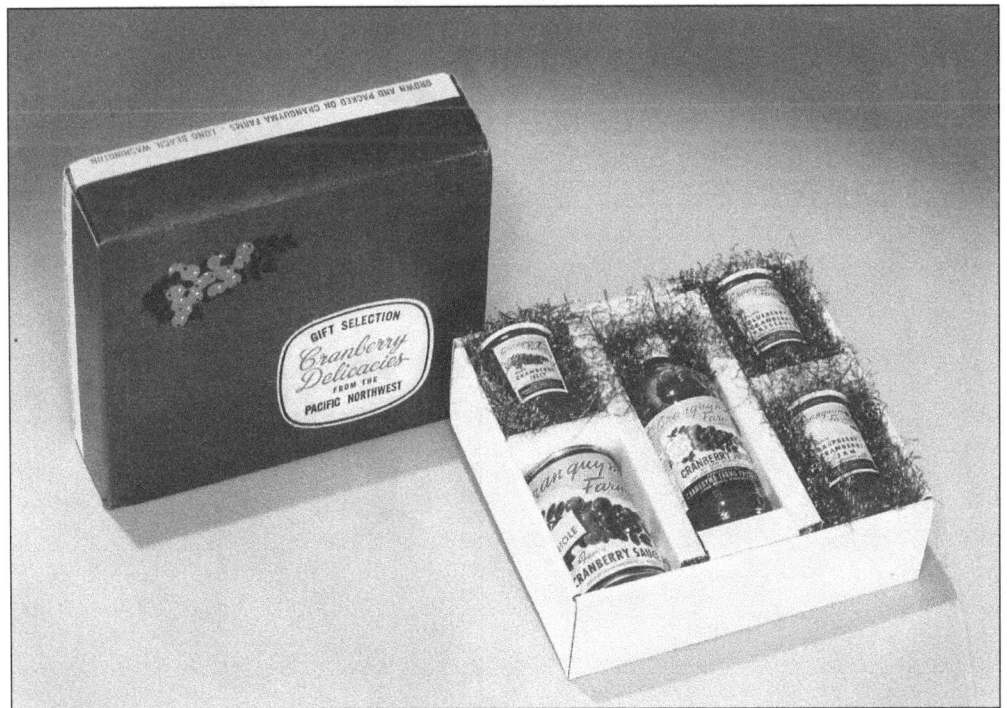

CRANBERRY JUICE. One of the earliest references to cranberry juice was in *The Compleat Cook's Guide*, published in 1683, but it would be many years until the product was made commercially. In 1930, the Ocean Spray Company was formed and, that same year, initially introduced canned cranberry sauce, followed by cranberry juice cocktail. Although the Prohibition era (1919–1933) was nearly at an end, it inadvertently affected the industry. Because cranberry products traditionally revolved around seasonal family holidays such as Thanksgiving and Christmas, Ocean Spray made a conscious choice not to associate itself with alcohol, so as not to risk alienating its largest consumer base. Ocean Spray did use the term *cocktail* to indicate the product had sugar added, but it is not clear if this influenced juice sales either one way or another. (Both, AMcPhail.)

THE HOLIDAY HURDLE. Some say the first Thanksgiving dinner at Plymouth in 1621 included wild cranberries but, according to the Smithsonian Institution, no documentation proves such a claim. Even so, because cranberry harvests happen in the late fall, the berries have long been associated with Thanksgiving. This connection was underscored in early years, before freezers were common household appliances and when cranberries were, of necessity, marketed fresh immediately after harvest. As harvesting and processing methods advanced, marketers worked hard to establish cranberries as a fruit to be eaten year-round. Today, of the 400 million pounds of cranberries consumed by Americans each year, only about 20 percent are eaten during Thanksgiving week, and of the seven to eight million barrels of cranberries that are produced in the United States each year, more than one-third of them are made into juice. (Both, MCrowley.)

CELEBRATING CRANBERRIES. For decades, festivals and parades on Washington's cranberry coast have provided opportunities for growers to draw attention to their berries and to the many products with which the fruit is associated—not only at harvesttime, but throughout the year. The Mist-Kist brand of the Grayland Cranberry Association, Inc. was featured in a 1939 Aberdeen parade and replicated in Grayland in the 1950s (above) in tribute to old-time celebrations. In Ilwaco, a traditional Fourth of July hayride takes modern-day celebrants on a tour of bogs adjacent to Black Lake. Not only growers but the entire coastal community takes pride in the bountiful bogs. (Above, JTweedy; below, AMcPhail.)

CRANBERRY BUTTONS. Each year since 1924, the Long Beach growers have celebrated their harvest with a gala community festival called the Cranberrian Fair. For a small fee to help defray expenses, buttons are sold as admission tokens and can be worn throughout the two-day celebration to assure entrance to all fair activities. Until growers had their own museum/gift shop/meeting room, the Cranberrian Fair was held in various locations. Nowadays, fair activities take place simultaneously at the Columbia Pacific Heritage Museum in Ilwaco and at Pacific Coast Cranberry Research Foundation headquarters on Pioneer Road, where walking tours of the bogs are offered in addition to traditional fair specialties. (Right, PCCRF; below, NStevens.)

WIDE-RANGING TASTE TREATS. Soon after its incorporation as a grower's cooperative in 1950, Ocean Spray began to explore ways to market the berries that had, up to that point, been so narrowly associated with Thanksgiving and Christmas. The cooperative's first product was jellied cranberry sauce, followed by its original Ocean Spray Cranberry Juice Cocktail. The Ocean Spray product line now includes a wide range of juice drinks, sauces, dried cranberries, and fresh fruits. Cranberries began to be appreciated as a fruit for all seasons and, as an added bonus, growers were relieved from the burden of doing their own marketing and distribution. (Both, AMcPhail.)

ASSOCIATED PRODUCTS. Ocean Spray marketing success has made the company itself a worldwide household word. The beverage cups and ball caps, tote bags and sweatshirts sporting the familiar Ocean Spray blue wave are worn proudly by growers and their families, and the slogans they feature—"As American as the 4th of July" and "Grower Owned Since 1930"—reinforce the story. The public has embraced the cranberry as something more than edible. It has become a subject for the decorative arts. Quilters, potters, jewelry makers, and all manner of craftsmen design with the cranberry in mind. Cranberries have come into their own. (Both, AMcPhail.)

JUSTIN AND HENRY. Actors Justin Hagan, 38 years old, and Henry Strozier, 71 years old, have been a part of Ocean Spray's advertising and public relations efforts for more than a decade. First featured in 2005, the campaign's comedic duo have captured the attention of the airways with more than 300 versions of the commercials. Their ads, filmed in a working cranberry bog in Massachusetts, were originally inspired by a picture of a couple of Ocean Spray growers standing in the bog during harvest. Although Justin and Henry have visited Oregon, Massachusetts, Rhode Island, Maine, Florida, and Wisconsin, they have yet to come to the Washington coast other than through the media. However, the Cranberry Museum was able to obtain not-quite-life-sized cardboard cutouts of them to use at a harvest celebration. Justin and Henry are shown here on the Ocean Spray bus that travels throughout the Eastern Seaboard as an extension of their "Straight from the Bog®" commercial campaign. The two "cranberry farmers" have been one of the most successful of Ocean Spray's advertising programs. (Both, AMcPhail.)

Five

A Thriving Community

UNWELCOME NEWS. In October 1991, the cranberry growers of western Washington received a letter from Arlen D. Davison, assistant dean of Washington State University. It began, "It is with deep regret that I must inform you that Washington State University will discontinue operation of its Long Beach Research and Extension Unit in mid-1992. . . . Discontinuing operation of the research and extension unit is the result of past and anticipated budget restriction." (AMcPhail.)

PROTESTERS GATHER. Immediately after WSU's announcement, the growers mobilized their forces and began to look at the possibilities. Their goal was to purchase the site and to keep the extension agent. They formulated a plan and asked to meet with WSU. The meeting was scheduled for December in Long Beach. In attendance were the Long Beach Growers Association, Grayland Growers Association, Ocean Spray Advisory Board members, representatives from the Markham and Long Beach Ocean Spray receiving stations, and Sid Snyder, Washington state senator. WSU was represented by Dean James Zuiches and Davison. By the time the meeting was over, they had agreed to work out a memorandum of understanding: WSU would sell the property to the growers and continue to support a research position and part-time secretary. The final contract was signed in February 1993. (Both, AMcPhail.)

PCCRF Begins. Growers formed a 501(c)(3), Pacific Coast Cranberry Foundation. It was "organized exclusively for education and scientific research." Since the bogs on the property were in poor shape, growers assigned themselves certain tasks. They weeded, sprayed, fertilized, and did the necessary jobs until funds became available through crop sales and membership dues to hire workers. For the first harvest, growers from Grayland and North Bay brought their Furford machines and many volunteers pitched in to help. Meanwhile, they renovated the bogs (two acres a year), built dikes, and enlarged the pond so that bogs could be water-harvested. It took 10 years to completely renovate all 11 acres of bogs on the demonstration farm, and some bogs have now been re-renovated with newer varieties. (Both, AMcPhail.)

INCLUSIVE MEMBERSHIP. Meanwhile, all West Coast farmers from Washington, Oregon, and British Columbia, as well as business and other associates, were encouraged to become PCCRF members. Until the first berries were harvested, monies from memberships were its primary source of income. Immediately, PCCRF purchased a piece of adjoining property, logged it for additional income, built a pesticide building, and put aside monies for remodeling the museum and other buildings. Many Washington growers got together for planned workdays and began to improve the property. Buildings were torn down, cleaned out, or eventually replaced thanks to grants from the Templin Foundation. Three pole buildings were added, and a much-needed meeting room was built onto the existing structure. (Both, AMcPhail.)

RENOVATION CONTINUES. After the old greenhouse was demolished and the pesticide building was constructed, the old mechanical shop was cleaned out and converted into a small, one-room museum. At first there was space for only a limited number of artifacts, but with a grant from the Ben B. Cheney Foundation, the PCCRF was gradually able to expand the museum space. The PCCRF named Melinda Crowley curator, and growers donated obsolete equipment and photographs to begin the collection. Charles Summers of South Bend was contracted to design the exhibit space and to provide high-quality photographic displays and informational signage. Gradually, the museum was expanded, and there was space enough for a larger gift shop to be added to the complex. (Both, AMcPhail.)

PCCRF Headquarters. Few people now remember that this building began as a drying shed and, by 1930, had become a shop and storage area for tools used on the farm. It is now headquarters for the Pacific Coast Cranberry Research Foundation and houses a museum and gift shop, a community hall, and an archival work area. This transformation has been gradual and has been made possible by generous grants from the Templin Foundation and from the Ben B. Cheney Foundation. Volunteer work parties of growers from North Bay, Grayland, and Long Beach spent many weekends doing the hard, physical labor of demolition, salvage, and reconstruction. The building continues to evolve as a well-used educational facility in addition to a busy venue for community activities. (PCCRF.)

CRANBERRY MUSEUM. Under the dedicated direction of curator Melinda Crowley, the PCCRF Museum has become a major attraction on the Peninsula. Oversized photographs and clear signage trace the history of the coastal cranberry bogs from earliest times to the present. Visitors can get a close-up view of tools and machinery specific to cranberry farming and may even be treated to a demonstration of the old bounce machine and learn how it can distinguish a good berry from a bad one. In addition to the interesting displays, the museum has been amassing an impressive archive of local cranberry information, which is available to researchers and scholars. Crowley, always on the lookout for more of the Washington coastal cranberry story, invites visitors to share their information and their memories. (Left, MCrowley; below, AMcPhail.)

HISTORIC ARTIFACTS. Bog shoes, cranberry scoops, bounce machines, and other unusual artifacts line the walls and peep forth from the nooks and crannies of this fascinating museum. With displays of historical and modern cranberry farming practices and harvest equipment, the exhibits are specifically related to the cranberry industry as it has developed on the Washington coast. Visitors come away with a greater understanding of the colorful industry that most had thought of only in terms of bright red berries floating across water-filled bogs. "I didn't know," or "I never realized," are the words that begin many conversations as visitors exit the exhibits and turn their attention to the gift shop beyond. The museum is open daily from April to mid-December, 10:00 a.m. to 5:00 p.m. (Both, NStevens.)

PCCRF Sweatshirt. The most popular item in the gift shop is the handsome PCCRF sweatshirt. It was the first merchandise that arrived before the not-quite-completed shop was up and running. Store manager Melinda Crowley put together a makeshift display area out on the porch, and, immediately, the sweatshirts began to sell. Designed by PCCRF member Ardell McPhail, the logo in the middle of the navy-blue shirt features a silver microscope with three red berries on a spray of cranberry vine with sun highlights in silver. The design is surrounded by "Pacific Coast Cranberry Research Foundation" in silver lettering. "Visitors love them," says Crowley. "We can't keep them in stock." (Both, AMcPhail.)

CRANBERRIES AND BEYOND. Not everything is edible, but everything in this gift shop is, in some way, cranberry related. "Or could be," says Crowley. Whether it is a painting or wearable art, cranberry-scented candles or soaps, or books about the ins and outs of the complex cranberry industry, visitors are likely to find it here. Cranberry sweatshirts, the shop's first for-sale item, have continued to be in demand. The refrigerator contains juice and cranberry ice cream—and one non-cranberry item. In a tribute to its longtime association with WSU, the PCCRF gift shop stocks Cougar Gold Cheese that, visitors are reminded, is the perfect accompaniment to a cranberry juice cocktail. From knickknacks and toy "Bog Bears" to cranberry-tinted glassware and wreaths of cranberry vines, there seems to be something for everyone. (Both, AMcPhail.)

USER-FRIENDLY. One of the many appealing exhibits at the Pacific Coast Cranberry Research Foundation is the self-guided walking tour through the cranberry bogs. Signs along the center dike road inform and direct the visitor along the way, and a complimentary brochure explains the various aspects of cranberry farming from growing and planting to water management and harvest. The walking tour provides a unique opportunity for a close look at a working cranberry bog, and depending upon the time of year, visitors may see field tests in progress or a harvest under way. Small colorful flags can be seen on various parts of the farm. They are used to mark plots where research on weeds, insects, or disease is being conducted. (Both, AMcPhail.)

Take a Walking Tour

of the

Pacific Coast Cranberry Research Foundation's

Demonstration Farm

The Tour is about 1/2 mile round-trip through the middle of the farm.

Use this informational brochure as your guide.

Pacific Coast Cranberry Research Foundation
Pioneer Road
Long Beach, Washington

VARIETY TRIALS. Visitors on the walking tour will pass an area of distinctive small square plots where variety trials take place. This is the area of the demonstration farm where new hybrids are evaluated for yield, keeping quality, and disease resistance. A good yield is 200 barrels per acre. Although there are over 200 varieties of cranberries, only 10 are produced commercially, the most popular West Coast varieties being Pilgrim and Stevens. At the conclusion of the walking tour, a "head-in-the-hole-board-for-three" beckons to would-be cranberry farmers to take a photograph (perhaps with their small child or dog) as a memento of their excursion. (Both, AMcPhail.)

THE MEETING ROOM. Built using grant money from the Templin Foundation, the meeting room is a new extension on the end of the old sorting shed/packinghouse. PCCRF members use the room for business meetings and social gatherings and, on occasion, have utilized the space to host concerts and book signings. Just as entrance to the museum and the self-guided walking tour around the bogs are free, there is no charge for community groups such as the local volunteer fire department or the oyster growers association using the space and only a nominal fee is charged for private parties. PCCRF's convenient location in the middle of the Peninsula makes it easily accessible. As seen above, the meeting room at PCCRF has been a popular venue from its beginning and was in use even before the inside walls were completed. (Both, AMcPhail.)

COMMUNITY GATHERING PLACE. The small kitchen adjacent to the meeting room is equipped with sinks, a stove, refrigerator, and ample counter space for practical use. It has also been approved as a venue for making the cranberry ice cream that is sold in the adjacent gift shop. Though meal preparation in the kitchen is possible, most groups who are planning to eat during or after their gatherings have their meals catered or host a potluck. Visitors and tour groups, whether seniors or children, are welcome to eat their box or bag lunches in the meeting room or on the large deck that runs the full length of the building. With its fine view of the demonstration bogs, the deck is a popular place for watching cranberry farming in progress. (Both, AMcPhail.)

A COASTAL CULTURE. All things cranberry permeate the Washington coastal culture. In both Grayland and Long Beach, roads carry the "cranberry" name, and both, as might be expected, take the traveler past acres of cranberry bogs, some of which are generations old. Even license plates reflect the cranberry culture. Businesses incorporate "cranberry" into their names and logos, menus in coastal restaurants feature cranberry recipes, and handcrafted products, from soap to candy, utilize the tart flavor and bright color of the berries to enhance their creations. The cranberry bogs themselves become subjects for local as well as for visiting artists, and at harvesttime, the entire community hosts fairs, parades, and celebrations in honor of the saucy red berries and the growers who produce them. (Above, AMcPhail; below, ADevlin.)

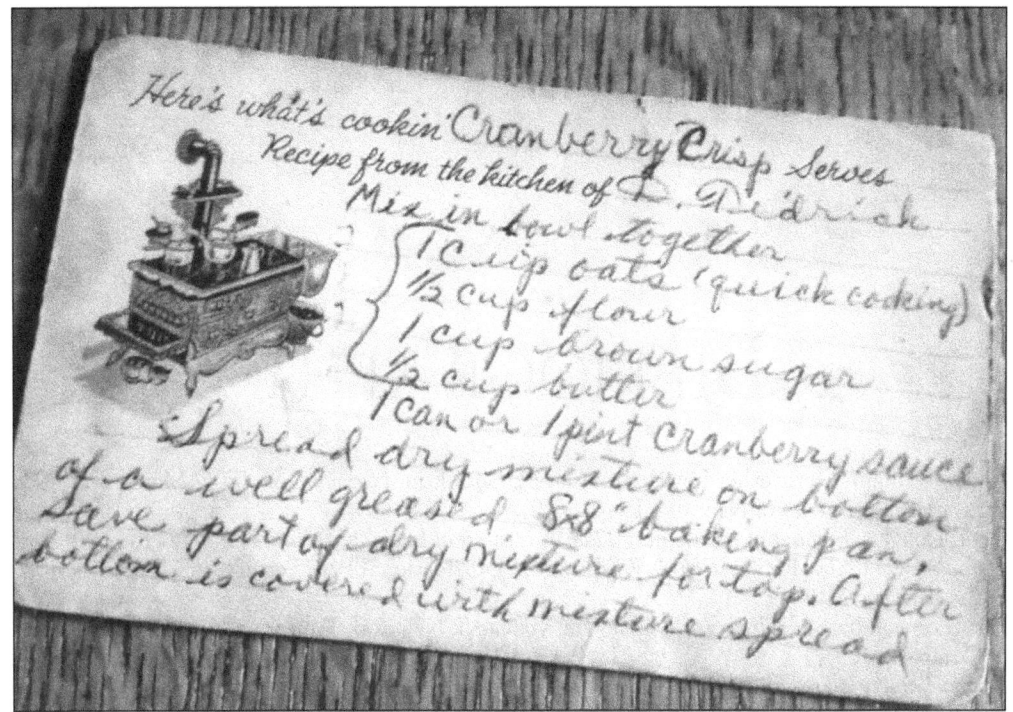

A CRANBERRY LEGACY. Cranberries are an integral part of coastal Washington's history. Old-timers still reminisce about the Ilwaco Railway and Navigation Co.'s Cranberry Station (1889–1930) located just south of the Cranberry Road crossing. There, growers deposited bags of fruit for the train to pick up on its way toward southern markets. Community festivities still mark each harvest, including the annual 130-year-old Cranberrian Festival. Precious family recipes are still shared from household to household and, during the winter holidays, strands of cranberries adorn community gathering places. According to long tradition, school groups get special invitations to tour the bogs and learn about the colorful, historic business of growing cranberries. "These are the cranberry farmers and supporters of the future," say growers. "We want them to be informed and to take pride in our cranberry legacy here on Washington's coast." (Above, LCrowley; below, AMcPhail.)

BIBLIOGRAPHY

Clark, William, Meriwether Lewis, and Gary E. Moulton. *November 2, 1805–March 22, 1806*. Lincoln: University of Nebraska Press, 2001.
Deur, Douglas. *Empires of the Turning Tide: A History of Lewis and Clark National Historical Park and the Columbia-Pacific Region*. Washington, DC: National Park Service, US Department of the Interior, 2016.
Hawthorne, Julian, ed. *History of Washington, Volume II*. New York: American Historical Publishing Company, 1893.
Ilwaco Tribune, various articles, 1925–1955.
Lloyd, Nancy. *Observing Our Peninsula's Past, Volume One*. Long Beach, WA: Chinook Observer, 2003.
———. *Observing Our Peninsula's Past, Volume Two*. Oysterville, WA: Oysterville Hand Print, 2006.
McDonald, Lucile. *Coast Country: A History of Southwest Washington*. Portland, OR: Binford and Mort, 1966.
Oesting, Marie, and Larry J. Weathers. *Oysterville Cemetery Sketches*. Washington, DC: M. Oesting, 1988.
Ruby, Robert H., and John Arthur Brown. *The Chinook Indians: Traders of the Lower Columbia River*. Norman: University of Oklahoma Press, 1988.
North Coast News (Ocean Shores, WA), various articles, 1992–1993.
Sou'wester. South Bend, WA: Pacific County Historical Society, 2018.
Turner, Frank. From Auld Lang Syne… (weekly column), *Ilwaco Tribune*, 1952–1958.
Williams, L.R. *Chinook by the Sea*. Portland, OR: Kilham Stationery and Printing Company, 1924.
———. *Our Pacific County*. Raymond, WA: Raymond Herald, 1930.
Wolf, Edward C. *A Tidewater Place*. Long Beach, WA: Willapa Alliance, 1993.

INDEX

Ben B. Cheney Foundation, 113, 115
Chabot, Anthony, 20, 21
Chabot, Robert, 21, 24
Clarke, Dr. J. Harold, 49
Copalis Crossing, 2, 21, 29
Cranguyma, 7, 8, 33, 38, 46, 49, 59, 60, 62, 100, 101
Crowley, D.J., 27, 33, 62
Crowley, Melinda, 7, 113, 118
Davison, Arlen D., 109, 110
Doughty, Charles C., 8, 9, 60
Elswa, Jonny Kit, 13
Funke, Guido, 96
Furford, Julius, 82, 83
Gane, Henry S., 26, 99
Grayland, 7, 9, 24, 29, 44, 46, 56, 58, 76, 82, 83, 85, 90, 95, 104, 110, 111, 115, 124
Grays Harbor County, 30
Hagan, Justin, 108
Holland, Dr. E.O., 26
Holman, Alexander, 34
Ilwaco, 28, 95, 96, 104, 105, 125, 126
Jacobson, Theo, 99
Lincoln, Abraham, 34
Lady Washington, 12
Long Beach, 2, 9, 16, 19, 20, 23, 25, 27–29, 34, 40, 50, 59, 60, 83, 90, 91, 93–95, 99, 100, 101, 105, 110, 115, 124, 126
Long Beach Research Station, 8, 9, 50, 109
Louisa Morrison, 14, 15

Marshall, John, 17
McPhail, Ardell, 7, 118
Meyers, Guy C., 49
Morgan, John, 17
Murakami, Ira, 62
Murakami, Jeff, 62
North Bay, 7, 33, 44, 46, 56, 58, 82, 90, 95, 111, 114, 115
Ocean Spray, 8, 9, 83, 90–95, 97, 100–102, 106–108, 110
Oysterville, 17–19, 28, 34, 95, 126, 127
Pacific County, 7, 18, 19, 28, 126
Patten, Dr. Kim, 6, 7, 9, 60, 61, 69
Paul, John Peter, 19
Pioneer Road, 28, 105
Shawa, Dr. Azmi Y., 9, 60, 61
Shoalwater Bay, 13–15, 18
Snyder, Senator Sid, 110
Strozier, Henry, 108
Summers, Charles, 113
Swan, James G., 13, 16
Templin Foundation, 112, 115, 122
Tjernberg, Chuck, 83
Tjernberg, Gwen, 83
Valentine, Vey, 62
Wakabayaski, Steich Dr., 62
Warness, Clarence, 85
Webb, John, 79
Zuiches, Dean James, 110

Discover Thousands of Local History Books
Featuring Millions of Vintage Images

Arcadia Publishing, the leading local history publisher in the United States, is committed to making history accessible and meaningful through publishing books that celebrate and preserve the heritage of America's people and places.

Find more books like this at
www.arcadiapublishing.com

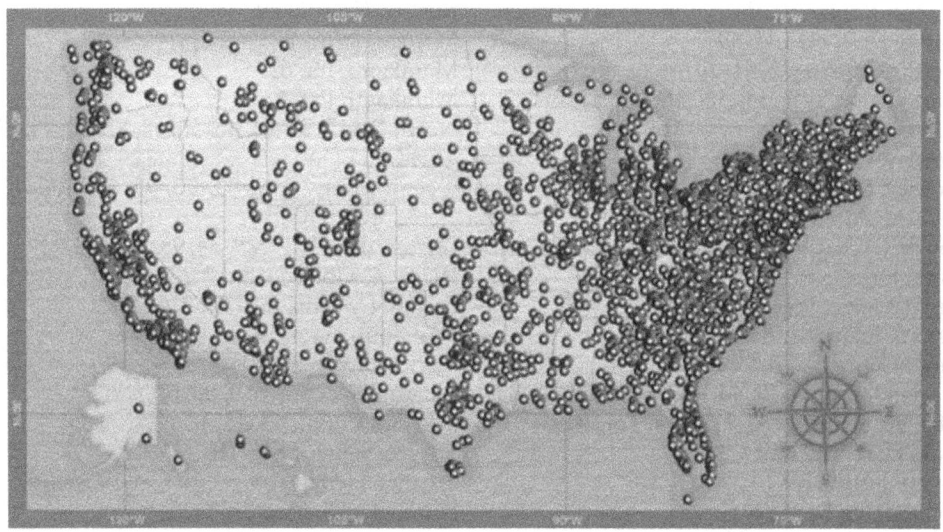

Search for your hometown history, your old stomping grounds, and even your favorite sports team.

Consistent with our mission to preserve history on a local level, this book was printed in South Carolina on American-made paper and manufactured entirely in the United States. Products carrying the accredited Forest Stewardship Council (FSC) label are printed on 100 percent FSC-certified paper.

www.ingramcontent.com/pod-product-compliance
Lightning Source LLC
Chambersburg PA
CBHW080911100426
42812CB00007B/2241